GERMAN JET GENESIS

GERMAN JET GENESIS

DAVID MASTERS

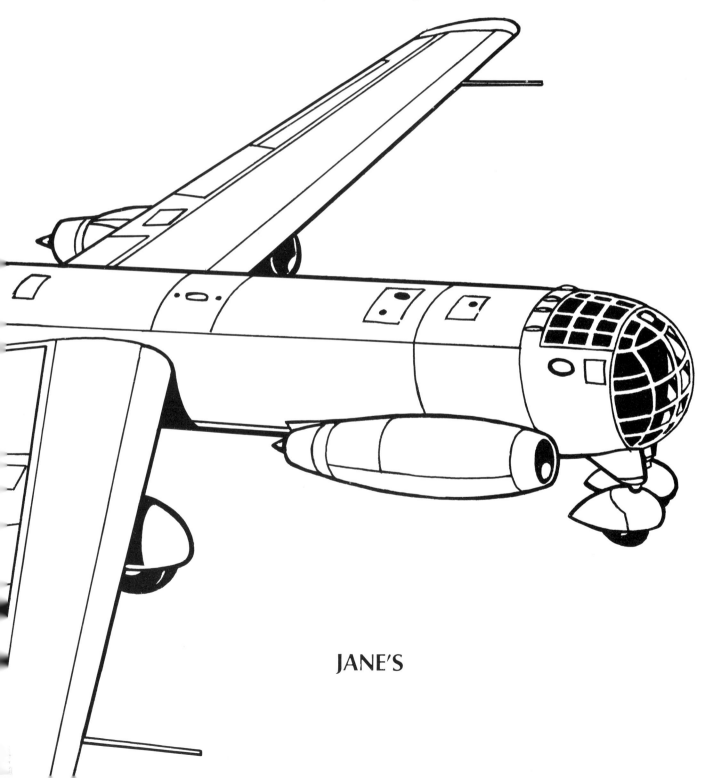

JANE'S

Copyright © David Masters 1982

First published in 1982 by
Jane's Publishing Company Limited
238 City Road, London EC1V 2PU, England

Distributed in Canada, the Philippines and the USA and its
dependencies by
Science Books International Inc
51 Sleeper Street, Boston
Massachusetts MA 02210

ISBN 0 86720 622 5

All rights are reserved. No part of this publication may be
reproduced, stored in a retrieval system, transmitted in any form by
any means electrical, mechanical or photocopied, recorded or
otherwise without prior permission of the publisher.

Computer typesetting by
Method Limited, Woodford Green, Essex
Printed in the United Kingdom by
Biddles Limited, Guildford, Surrey

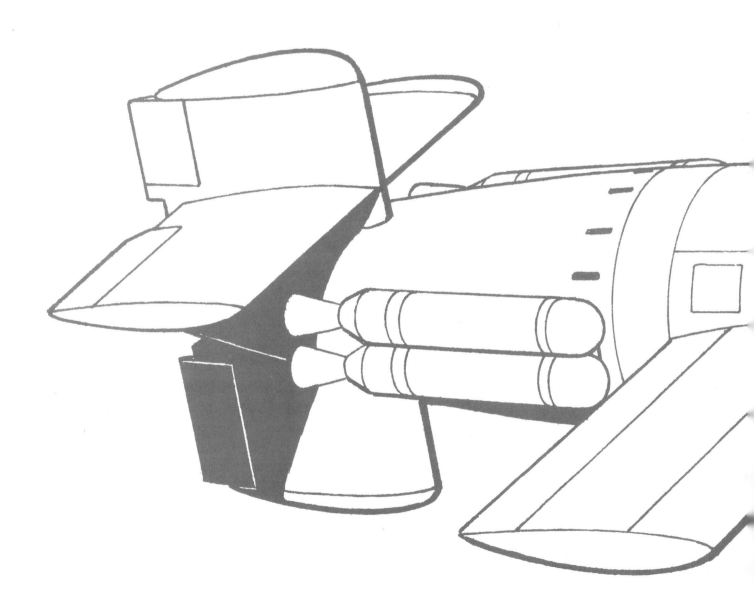

Contents

Introduction	9
Chronology	13
Alphabetical list of types	17
Appendix	139

Introduction

On September 30, 1929, a little glider-like aircraft designated Rak-1 took to the air amidst a cloud of smoke to become the first piloted aircraft to take off and fly under the sole power of a reaction powerplant, in this case a battery of simple solid-fuel rockets. At the controls was Fritz von Opel, who, together with Max Valier of the German *Verein für Raumschiffahrt* (VfR, Society for Space Navigation), performed and prompted a number of experiments with rocket aircraft in a publicity campaign to help finance the private development of liquid-fuel rocket motors. Valier's initiative proved to be a false start on the road to jet-powered flight, however, and it was not until 1937 that a serious attempt at designing a rocket aircraft was made again in Germany.

That year saw the launch of two projects: a private venture by Ernst Heinkel and his designers based on a Walter liquid-fuel rocket motor, and an official project initiated by the *Reichsluftfahrtministerium* (RLM, State Ministry of Aviation) under the control of Dr Alexander Lippisch, which was code-named *Projekt* X. Lippisch worked steadily on *Projekt* X through the war until it culminated in 1944 in the world's first rocket-powered fighter, the Messerschmitt Me 163B *Komet*. Ernst Heinkel received little official encouragement, however, and his project fizzled out after a short series of tests with the diminutive He 176 research aircraft.

But despite the very high speeds that they made possible, rockets had a number of inherent drawbacks, principally their limited endurance and unreliability, and the lack of any means of regulating their thrust. So it was that rocket development was paralleled almost from the beginning by work on the gas-turbine jet engine, or turbojet. Heinkel had foreseen the possibilities of this type of engine in 1936, when he employed two talented engineers to develop their own jet-powered aircraft. The two men, Pabst von Ohain and Max Hahn, worked on their project alongside the He 176 team until in August 1939 their He 178 took to the air for a short powered flight, making it the world's first turbojet-powered aircraft to fly.

From 1939 German jet and rocket aircraft research followed a multitude of paths as the necessary liquid-fuelled rockets and turbojet engines – as well as pulsejets, ramjets and turboprops – were developed. This book includes some two hundred designated or named jet, mixed propeller/jet and rocket aircraft designs dating from 1929 up to the end of the Second World War. These begot another hundred variants with such strong individual characteristics that they could almost have been treated as separate projects. By far the majority of the designs incorporated turbojet engines, mostly as the sole form of propulsion but sometimes combined with piston engines to improve low-speed endurance, or with rocket motors for use as take-off or short-duration performance boosters. In all, over 130 projects were based on turbojet engines, and of these less than 10 per cent had supplementary piston engines.

The Second World War naturally gave a sense of great urgency to the development of jet and rocket aircraft, and the imagination of the German designers seems to have been given free rein. The result was a proliferation of very advanced and truly amazing designs. But so new and revolutionary were jet and rocket propulsion that comparatively few of these projects actually made it into the air. The Germans nevertheless put the Allied effort – one operational type, the Gloster Meteor – firmly in the shade, getting into flight test fully 25 jet or rocket types, of which five were actually delivered to *Luftwaffe* units: the Arado Ar 234, Bachem Ba 349, Heinkel He 162, Messerschmitt Me 163 and Messerschmitt Me 262. Of the lesser types, a half dozen or so were in the process of being built when the war ended, while some two dozen others had achieved an advanced stage of design before they were abandoned or overtaken by events. Of the other 150 designs described, most were cancelled at the drawing-board stage. Nevertheless, their descriptions still make interesting reading, if only as examples of the extraordinary creativity and ingenuity of Germany's wartime aircraft designers.

While the gas turbine soon established its all-round superiority, the pressing needs of war forced the

Germans to look elsewhere for simpler, cheaper jet powerplants. Attention therefore turned to liquid and solid-fuel rockets, pulsejets and ramjets, even though all of these had severe limitations when compared with the gas turbine. Liquid-fuel rockets, though incorporated as the prime power source in some 30 of the aircraft described here, suffered from short endurance and poor controllability, and a serious danger of fire or explosion, such was the instability of their fuels. Pulsejets were very limited in output, inflicted serious acoustic damage on airframes, and suffered a severe drop in power as height was gained, making them suitable only for use in low-level roles. Some eight designs were to have ramjets as their main power source, and at first sight this type of powerplant appears to be the best compromise between simplicity and performance. The power that a ramjet can produce is almost limitless as long as enough fuel and air can be delivered to it, and it is by far the simplest type of jet engine, being little more than an internally contoured tube into which air is rammed through the front orifice, compressed, mixed with fuel and then ignited to exhaust from the rear. However, on the other side of the coin are an enormous thirst for fuel and a reluctance to operate at speeds below about 200mph (320km/hr). The latter limitation meant that ramjet-powered aircraft had to have some form of auxiliary propulsion for take-off and acceleration to ramjet operating speed. But for all their shortcomings, these powerplants opened up for the German designers an array of possibilities which they exploited with an energy and inventiveness that staggered the victorious Allies. Containing descriptions of aircraft designed for anything from submarine spotting to intercontinental bombing, this book amply illustrates the versatility of the reaction-powered aircraft, even at an early stage in its development.

Unfortunately, their highly secret military nature and the chaos that afflicted Germany at the end of the war have meant that most of the documents so vital to an analysis of this fascinating period of aviation history have been lost or destroyed or are otherwise inaccessible. What information does survive is often very vague and sometimes contradictory when different references are compared. This problem particularly afflicts some of the more revolutionary projects, such as the Schriever flying discs, the traces of which either were destroyed at the approach of Allied forces or disappeared behind the Iron Curtain. Indeed, it is quite possible that some descriptions are now a mixture of fact and legend, and the reader is asked to approach some of the less fully described or more obviously revolutionary projects with this in mind. I have been at pains to include in this book as much well substantiated information as possible, but was equally concerned not to include fiction. Thus whenever the references were conflicting or vague, the descriptions have been worded so as to warn the reader that doubts exist.

The data tables accompanying the entries for the better documented projects contain an indication of their ultimate status, showing whether they remained designs only, entered construction or flight test, or became operational. In the case of those projects that remained on the drawing board, all performance figures are designers' estimates.

David Masters
January 1982

Chronology of German jet and rocket-powered aircraft development

1927 *Verein für Raumschiffahrt* (VfR, Society for Space Navigation) founded in Germany. Originators and first members include Prof Hermann Oberth, Klaus Reidel, Rudolf Nebel, Max Valier, Willy Ley and the young Wernher von Braun.

1928 Max Valier of VfR initiates a series of public displays of rocket propulsion, resulting in a promise of financial support from Fritz von Opel, the car manufacturer.
May Opel Rak-2 rocket-propelled car achieves 143mph (230km/hr) on the Avus race track near Berlin.
June 11 First flight of a manned aircraft powered solely by rockets, by the Lippisch-Sander *Ente* glider piloted by Fritz Stamer at Wasserkuppe. It was powered by two Sander powder rockets each generating 44lb (20kg) thrust for 30sec, and take-off was assisted by a bungee (rubber) catapult.

1929 September 30 At a location near Frankfurt the Opel-Hatry Rak-1 glider, powered by 16 Sander powder rockets developing a total of 882lb (400kg) thrust and piloted by Fritz von Opel, becomes the first manned aircraft to take off and fly on reaction power alone.

1930 May 4 First flight of the rocket-powered Espenlaub/Sohldenhoff E.15 tailless glider, which took off from a site near Bremerhaven under the control of Gottlob Espenlaub.
May 17 Max Valier killed while testing a liquid-fuel rocket unit.
Dipl Ing Paul Schmidt applies for a patent for his pulsejet propulsion unit, later known as the *Schmidt-Rohr*.

1935 Hans-Joachim Pabst von Ohain applies for his first patent, entitled "Method and apparatus for producing airflow for aircraft propulsion" (patent granted in 1937).
Start of development work on liquid-fuel rocket motors by Hellmuth Walter.

1936 April 15: Pabst von Ohain and Dipl Ing Max Hahn begin development work on a turbojet aircraft engine at the Heinkel plant at Marienehe.
German Research Institute for Rocket Flight established.
March First flight test of a liquid-fuel rocket motor developed by von Braun and producing 286lb (130kg) thrust in a He 112. The aircraft exploded but the pilot, Erich Warsitz, thrown clear.
April First successful flight by a He 112 with a built-in von Braun liquid-fuel rocket motor. Pilot is Erich Warsitz.
First bench tests of a Walter liquid-fuel rocket motor at Kiel.

1937 February Walter liquid-fuel rocket motor developing 298lb (135kg) thrust for 45sec flight-tested in a He 72 *Kadett* biplane.
March First bench tests of von Ohain's HeS 2A turbojet, developing just over 286lb (130kg) st, at Marienehe.
April Walter liquid-fuel rocket motor developing 595lb (270kg) thrust for 30sec flight-tested in a FW 56 *Stösser*.
Dr Hans Regner at Heinkel begins design work on the He 176 rocket-powered research aircraft.
Projekt X, the development of a rocket-powered aircraft designed by Dr Alexander Lippisch at DFS, initiated by RLM. Subsequently designated 8-194.
Summer He 112 piston-engined fighter testbed fitted with von Braun liquid-fuel rocket motor takes off and flies solely under rocket power.
Start of serious experiments with ramjets.

1938 Summer HeS 3B turbojet, rated at 992lb (450kg) st, test-flown in a He 118.
Autumn RLM orders a jet fighter from Messerschmitt to be powered by two projected BMW turbojets. Design becomes P.1065.
Dr Ing Günther Dietrich at Argus/Berlin discovers pulsejet principle independently of Paul Schmidt.

1939 January 4 RLM circulates a top-secret discussion paper entitled "Preliminary technical guidelines for high-speed fighters with turbojet propulsion".
February RLM contracts BMW to develop a turbojet engine of 1,323lb (600kg) st. Design team headed by Dr Hermann Oestrich.
First bench tests with an experimental Junkers turbojet engine, the forerunner of the Jumo 004.
Spring Work begins on BMW P.3302 turbojet, which eventually leads to BMW 003.
June 7 Messerschmitt submits a twin-turbojet fighter study, the P.1065, to the RLM.
June 20 First flight of the He 176 research aircraft powered by one Walter HWK R.I-203 unit of 1,102lb (500kg) thrust at Peenemünde. It is the first manned aircraft designed to be powered by a liquid-fuel rocket engine. Pilot is Erich Warsitz.
July Start of detailed development work on the Jumo 109-004 turbojet.
August 27 First flight of the He 178, powered by a Heinkel-Ohain HeS 3B of 992lb (450kg) st and piloted by Erich Warsitz. It is the world's first pure jet aircraft.
October 16 Start of ground tests of the DFS 194 prototype fitted with the Walter TP-1 liquid-fuel rocket motor.
December Messerschmitt P.1065 mock-up inspected by RLM officials.

1940 February First flight tests of the DFS 194 in unpowered glider form and piloted by Heini Dittmar.
February Attempts made on RLM instructions to co-ordinate the work of Paul Schmidt and Günther Dietrich of Argus in search of an effective pulsejet.
March 1 Messerschmitt P.1065 approved by RLM and allocated the designation Me 262. Development contract awarded to Messerschmitt.
August First flight of the DFS 194 under rocket power, at Peenemünde-West. Pilot is Heini Dittmar.
September 11 First flight of the He 280 jet fighter prototype in glider form.
October Start of Jumo 004 bench tests.
October RLM issues a specification to Arado for a high-altitude, medium-range reconnaissance aircraft to be powered by two of the BMW or Jumo turbojets then under development.

1941 February Arado design team headed by Dipl Ing Walter Blume and Ing Hans Rebeski submits the E.370 turbojet-powered reconnaissance project, which is accepted by the RLM as the Arado 8-234.
March First flight of the Me 163 prototype in glider form and piloted by Heini Dittmar.
March 20 First flight of the He 280 jet fighter prototype powered by two HeS 8A turbojets developing 1,653lb (750kg) st each and piloted by Fritz Schäfer.
April 18 First flight of the Me 262 prototype, powered by a single Jumo 210G piston engine (due to the unavailability of the turbojets) and piloted by Fritz Wendel.
April 30 First flight tests of an Argus pulsejet of 265lb (120kg) st, fitted under a Go 145 biplane. Promising results lead to subsequent flight tests of improved Argus pulsejets fitted to Do 217, Ju 88 and Bf 110 testbeds.
August 13 First flight of a Me 163A prototype, powered by HWK R.II-203 liquid-fuel rocket motor of 1,653lb (750kg) st, at Peenemünde. Pilot is Heini Dittmar.
October 2 Third Me 163A prototype, piloted by Heini Dittmar, attains 623.85 mph (1,004km/hr) at Peenemünde, a new world speed record that remains secret until the end of hostilities.
In response to an RLM request for a new interceptor fighter Dr Eugene Sänger proposes a ramjet as the sole power unit. Development and construction work on large ramjets authorised to go ahead at DFS.
December RLM initiates the *Amerika-Bomber* programme.

1942 March 7 First flight tests of a Sänger ramjet mounted above a Do 17Z. Pilot is Paul Spremberg, accompanied by Dr Sänger.
March 25 Unsuccessful flight test of the first Me 262 prototype powered by two early BMW 003 turbojets of 1,015lb (460kg) st each and a central Jumo 210G piston engine. Pilot is Fritz Wendel.
March First taxiing trials of the Ar 234A prototype powered by two pre-production Jumo 004 turbojets.
July 18 First flight of the Me 262 prototype powered by two Jumo 004A turbojets of 1,850lb (840kg) st each. Pilot is Fritz Wendel.
Dr O. Pabst at Focke-Wulf begins development of subsonic ramjets.
Development work on the HeS 30, BMW 002 and DB 007 turbojets is stopped.
RLM issues a design request for a single-turbojet fighter to selected German manufacturers.
December First bench tests of the redesigned BMW P.3302 turbojet, the future BMW 109-003A.

1943 March 27 He 280 officially turned down as a fighter by the RLM.
May Walter HWK 109-509A-0 liquid rocket motor pre-production series is completed.
June Dipl Ing Hans Wocke at Junkers begins project work on the radical turbojet-powered

bomber with negative wing sweep that eventually becomes Ju 287.
July 30 First flight of the Ar 234A prototype powered by two pre-production Jumo 004A turbojets developing 1,850lb (840kg) st each. Pilot is *Flugkapitän* Selle.
August Me 262 ordered into series production.
September First flight of Me 328 glider prototype.
Initial flight tests of the first DFS 228 prototype in glider form at Hörschingen and Rechlin. The aircraft is released from a Do 217K carrier.
October First flight tests of the BMW 003A turbojet mounted under a Ju 88 testbed.
Official decision to concentrate on jet propulsion for all future fighters and bombers.

1944 February Start of Jumo 004B *Orkan* series production.
First bench tests of the Heinkel HeS 011 turbojet, though full 2,866lb (1,300kg) st output is not achieved until early 1945.
March RLM orders prototype development of the Ju 287 turbojet-powered bomber by Junkers design team headed by Dipl Ing Ernst Zindel.
March First flight tests of Me 328 prototype powered by two Argus As 014 pulsejets developing 661lb (300kg) st each.
April First flight tests of the Horten Ho IX glider.
April First Me 262A turbojet fighters delivered to the Luftwaffe.
May First operational use of the Me 262A *Schwalbe*.
May RLM issues requirement for a target-defence interceptor.
July First operational use of the Me 163B *Komet*.
August 2 First operational use of the Ar 234A *Blitz*.
August 16 Initial flight of the first Ju 287 prototype, powered by four Jumo 004 turbojets. Pilot is Siegfried Holzbauer.
August BMW 109-003A *Sturm* accepted for series production.
August Dipl Ing Erich Bachem and Willy Fiedler prepare their initial design for a radical vertical take-off target-defence interceptor, the BP 20.
August Start of Walter HWK 109-509A-1 rocket motor series production.
August RLM issues Emergency Fighter requirement, specifying one HeS 011A turbojet as powerplant, to Blohm und Voss, Focke-Wulf, Heinkel, Junkers and Messerschmitt.
September 8 Basic specification for a *Volksjäger* drawn up by the RLM and issued to Arado, Blohm und Voss, Fieseler, Focke-Wulf, Heinkel, Junkers and Messerschmitt.
September 23 Despite the obvious superiority of the Blohm und Voss proposal the Heinkel P.1073 is declared the winner of the *Volksjäger* competition and ordered under the designation He 500 (later changed to He 162).
September First operational use of the Ar 234B *Blitz*.
September Bachem BP 20 proposal accepted for development by the RLM and allocated the designation Ba 349 *Natter*.
September First flight of the *Reichenberg* IV (manned Fi 103) in glider form.
October First flight of the *Reichenberg* IV powered by an Argus As 014 pulsejet.
October Large Pabst ramjet tested in flight on a Do 217 carrier.
RLM issues an official requirement for the development of a ramjet-powered day interceptor fighter.
November RLM issues *Miniaturjäger* specification for a simple jet-propelled mass-production day fighter.
December 6 Initial flight of the first He 162 *Salamander* prototype, at Schwechat. Pilot is Gotthold Peter.
December 18 First vertical launch of an unmanned Bachem Ba 349 *Natter*.

1945 late January Formation of *E-Kdo* 162, the *Volksjäger* service evaluation and trials unit.
February 6 First He 162As delivered to an operational Luftwaffe unit, for training purposes.
February 24 First almost intact Ar 234B falls into Allied hands after being forced down by P-47 Thunderbolts.
February 25 First manned flight test of the vertical take-off Ba 349 rocket-powered target-defence interceptor kills the pilot, Lothar Siebert.
March Focke-Wulf P.VI/I *Huckebein* declared winner of the Emergency Fighter competition and is immediately ordered into development and production as the Ta 183.
March Three successful launches of manned Ba 349As flown by other volunteer pilots; *Natter* accepted for operational service.
April First operational Ba 349A *Natter* unit formed and deployed, though the aircraft is never used in action.
April First reported (but unconfirmed) operational use of the He 162A.
April Henschel Hs 132 single-seat turbojet-powered dive bomber/attack aircraft prototype ready for flight tests. Arrival of Soviet forces means that it is never flown.
June onwards Last eight HeS 011A-0 pre-production turbojets completed on American orders and shipped to the USA, subsequently undergoing US Navy bench tests at Trenton.
October Dr Ing Oestrich and a group of other German turbojet engineers begin design work at Rickenbach on a new turbojet for the French, the ATAR 101.

1945- Flight tests of captured German jet aircraft in the
46 United Kingdom. Flt Lt Marks killed while flying a He 162.
1946 October Completion of dismantling and transfer of all aircraft plants, test facilities and staff in the Soviet Zone of Germany to the Soviet Union.
1948 Spring First glide tests of the DFS 346 prototype in the USSR.
1949 First flight tests of the rocket-powered DFS 346 prototype in the USSR.

Arado Ar 234 (E.370) Blitz

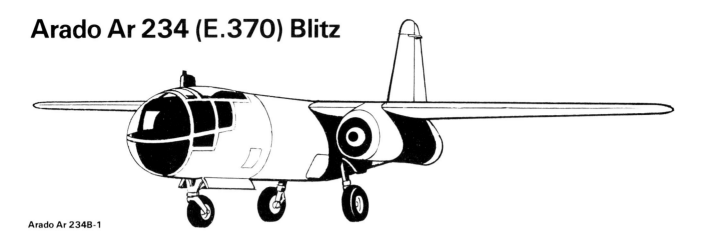

Arado Ar 234B-1

By the end of 1940 it had become clear to the Luftwaffe that it would need a very advanced bomber to continue its bombing campaign against Britain. To this end the Arado Ar 234, a shoulder-wing, twin-jet monoplane with very clean lines, was developed. However, although the airframe was ready late in 1941, development of the turbojet engines took longer than expected and the first Ar 234 prototype did not fly until July 30, 1943.

Originally projected under the Arado designation E.370, the Ar 234 was to have had a retractable undercarriage comprising nine pairs of wheels. They were very narrowly tracked, necessitating the support of a retractable skid beneath each of the engines. This arrangement ultimately gave way to a jettisonable tricycle trolley for take-off, and a retractable belly skid, supplemented by the smaller underwing skids, for landing. This method was used on the first eight "V" (*Versuchs*, experimental) examples but was discontinued with the V9, which was given a tricycle undercarriage retracting into the fuselage. This remained standard on subsequent versions, with the bombs mounted externally beneath the inner wing sections.

The experimental programme was very extensive, with many variations on the basic theme. Some examples were fitted with as many as four turbojets

Ar 234 V3 landing on its belly skid (*via Alex Vanags*)

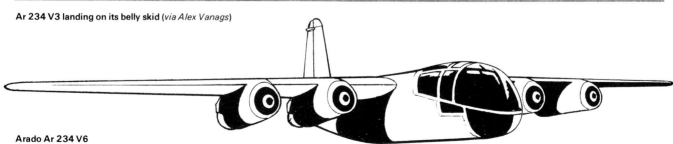

Arado Ar 234 V6

Ar 234B takes off with rocket assistance (*via Alex Vanags*)

Inset: **Ar 234 V8 mounted on a take-off trolley** (*via Alex Vanags*)

beneath the wings; one such was the V6, which consequently had its maximum speed boosted to over 530mph (855km/hr). But despite these efforts only the single-seat B-1 and B-2, named *Blitz* (Lightning), saw operational service. Both were fitted with a pair of Jumo 004B turbojet engines, the B-1 being a reconnaissance aircraft and the B-2 a bomber. The B-1 first flew in December 1943 and entered production in June 1944. The Ar 234A was first operational over Normandy in August 1944, followed a month later by the Ar 234B. Both variants had a defensive armament of two 20mm cannon mounted in the tail and aimed with an ingenious rear-facing periscope by the pilot, who was the sole crew member. In all, some 210 Bs were completed in a building programme which would have run into several hundreds more if the war had continued.

The more powerful Ar 234C (four BMW 003As in paired nacelles) and Ar 234D (two HeS 011As) were also planned. An Ar 234C did actually fly as a prototype in the early months of 1944, but neither series entered service. Another version, the P series, was envisaged as a night fighter, equipped with radar and pressurised cabins for the crew of two or three. This remained a project only, as did an even greater extension of the design, the R-series reconnaissance variant. This exotic version was expected to be powered by a single Walter rocket motor of 4,400lb (2,000kg) thrust which was to be used only to gain operational height after the aircraft had been released by a towing Heinkel He 177 at around 26,000ft (8,000m). This mode of operation was also envisaged for the DFS 228 (see page 45).

Also worthy of mention is a version of the Ar 234 which was to use the new Heinkel HeS 021 propeller-turbine engine, but again this remained only a project.

Even as the war drew to a close the Ar 234 was being earmarked for greater things. The minutes of an aircraft development programme meeting on November 21-22, 1944, gave the highest priority to four types – the He 162, Me 262, Do 335 and the Ar 234 – and there was a proposal to build a swept-wing version similar in appearance to the Arado E.560 (see page 21).

Arado Ar 234C

Ar 234 data

Variant/role	Crew	Powerplant	Weight lb (kg)	Max speed at height mph, ft (km/hr, m)	Range miles (km)	Ceiling ft (m)	Armament
B-1 reconnaissance	1	2×Jumo 004B, each 1,984lb (900kg) st	—	485 at 19,690 (780 at 6,000)	1,200 (1,930)	37,730 (11,500)	2× MG 151/20 20mm cannon
B-2 bomber	1	2×Jumo 004B, each 1,984lb (900kg) st	20,280 (9,280)	472 at 19,690 (760 at 6,000)	1,000 (1,610)	37,750 (11,500)	2×MG 151/20 20mm cannon, 2,205lb (1,000kg) bombs
C-1 bomber	1	4×BMW 003A, each 1,764lb (800kg) st	21,690 (9,840)	540 at 19,690 (870 at 6,000)	910 (1,465)	37,730 (11,500)	2×MG 151/20 20mm cannon, 2,205lb bombs
C-2 bomber	1	4×BMW 003A, each 1,764lb (800kg) st	22,250 (10,090)	553 at 19,690 (890 at 6,000)	1,000 (1,610)	37,730 (11,500)	—
C-3 bomber	1	4×BMW 003A, each 1,764lb (800kg) st	24,360 (11,050)	553 at 19,690 (890 at 6,000)	765 (1,230)	37,730 (11,500)	2×MG 151/20 20mm cannon, 2,205lb bombs
C-4 reconnaissance	1	4×BMW 003A, each 1,764lb (800kg) st	20,000 (9,070)	547 at 19,690 (880 at 6,000)	—	37,730 (11,500)	2/4×MG 151/20 20mm cannon
C-5 bomber	2	4×BMW 003A, each 1,764lb (800kg) st	—	—	—	37,730 (11,500)	2×MG 151/20 20mm cannon, 2,205lb bombs
C-6 reconnaissance	1	4×BMW 003A, each 1,764lb (800kg) st	—	—	—	37,730 (11,500)	—
C-7 night fighter	2	4×HeS 011A, each 2,866lb (1,300kg) st	24,360 (11,050)	—	—	—	—
C-8 bomber	1	2×Jumo 004D, each 2,205lb (1,000kg) st	21,605 (9,800)	472 at 19,690 (760 at 6,000)	—	—	2×MG 151/20 20mm cannon, 2,205lb
D-1 reconnaissance	1	2×HeS 011A, each 2,866lb (1,300kg) st	—	—	—	—	—
D-2 bomber	1	2×HeS 011A, each 2,866lb (1,300kg) st	—	—	—	—	—
P-1 night fighter	2	4×BMW 003A, each 1,764lb (800kg) st	25,800 (11,700)	528 at 19,690 (850 at 6,000)	700 (1,125)	—	1×MG 151/20 20mm cannon, 1×MK 108 30mm cannon
P-2 night fighter	2	4×BMW 003A, each 1,764lb (800kg) st	25,800 (11,700)	528 at 19,690 (850 at 6,000)	700 (1,125)	—	1×MG 151/20 20mm cannon, 1×MK 108 30mm cannon
P-3 night fighter	2	2×HeS 011A, each 2,866lb (1,300kg) st	23,470 (10,645)	510 at 19,690 (820 at 6,000)	1,056 (1,700)	—	2×MG 151/20 20mm cannon, 2×MK 108 30mm cannon
P-4 night fighter	2	2×Jumo 004D, each 2,205lb (1,000kg) st	23,150 (10,500)	441 at 19,690 (710 at 6,000)	1,070 (1,720)	—	2×MG 151/20 20mm cannon, 2×MK 108 30mm cannon
P-5 night fighter	3	2×HeS 011A, each 2,866lb (1,300kg) st	—	—	—	—	4×MK 108 30mm cannon

All wing dimensions the same at 47ft 4¼in (14.44m) span and 298.16ft² (27.7m²) area. All lengths the same at 41ft 6in (12.66m), except P variants at 43ft 6in (13.26m).

Arado Ar 240A (with Jumo 004)

The Ar 240 was a multi-purpose low-wing monoplane with twin fins and rudders. Its two DB 601E piston engines were in long nacelles which projected well behind the wing trailing edges. One version was to be fitted with a Jumo 004B turbojet beneath the fuselage for use in emergency to boost the maximum speed to 475mph at 19,690ft (765km/hr at 6,000m). However, although some conventionally powered examples entered operational service, the Ar 240 was still very much under development when the *Reichsluftfahrtministerium* (RLM, State Ministry of Aviation) eventually decided against quantity production.

Arado E.375

Arado design projects were usually given an "E" (*Entwicklungs*, development) prefix and – as in the case of the E.370, which became the Ar 234 when it was decided to go into development – redesignated in the "Ar" series if the concept got off the drawing board. The E.375 was not developed and consequently there is little information about this project, except that it was intended as a fast bomber, possibly an eventual replacement for the Ar 234. It was to be fitted with two of the very powerful 11-stage axial-flow Jumo 012 turbojets, expected to develop approximately 6,530lb (2,960kg) st each.

Arado E.381

The design work on this unusual midget rocket interceptor was well advanced by the end of 1944. It seems to have been envisaged in at least two forms which were similar in basic layout and purpose but with different armament.

The first and more highly developed version had a very low profile, with twin fins and a prone pilot; this layout was chosen specifically to allow easy stowage beneath an Ar 234C carrier aircraft. It was to be similar in appearance to the BV 40 glider fighter, with shoulder wings and armament housed in the wing roots. But unlike the BV 40, which was to be towed into action, the E.381 was to have a single Walter HWK 109-509B rocket motor mounted centrally to power it into the attack after it had been released by the carrier aircraft. (At one point rocket or pulsejet power were also considered for the BV 40 in a last-ditch effort to save the project before the RLM finally cancelled it.)

The interception was intended to be short and sharp once a target had been selected, and an armament of six small missiles located in rows of three in separate wing-root tubes was regarded as sufficient to inflict the desired damage. The return journey would be made in a long glide, followed by a landing on a central retractable skid.

The second version had a deeper and shorter fuselage (16ft 3in, 4.95m) and a high mid-wing layout, and was to be powered by a Walter 109-509A-2 unit rated at 3,750lb (1,700kg) thrust. About a quarter of the way back from the nose the fuselage deepened in the form of a hump which extended to the tail, housing a single MK 108 30mm cannon with 45 rounds of ammunition. The bulkier fuselage was accommodated beneath the carrier aircraft by means of a recess under the Ar 234C's mid-fuselage.

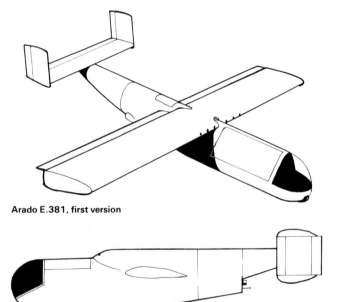

Arado E.381, first version

Arado E.381, second version

Arado E.381 data*

Role	Single-seat rocket interceptor
Ultimate status	Design
Powerplant	One Walter 109-509B rocket motor, 772lb (350kg) thrust
Maximum speed	560mph at 26,250ft (900km/hr at 8,000m)
Weight	3,307lb (1,500kg) loaded
Span	16ft 5in (5.00m)
Length	18ft 8½in (5.70m)
Wing area	59.2ft² (5.50m²)

*Shoulder-wing version with rocket armament.

Arado E.385

Like the E.375, the E.385 was intended for use as a fast bomber. It was also to be powered by two turbojet engines of greater power, BMW 018 12-stage axial-flow turbojets developing a remarkable 7,496lb (3,400kg) st each. The project's fate is not known, and it was presumably scrapped or overtaken by events in 1945.

Arado E.395

Another late project, the E.395 was designed as a bomber and reconnaissance aircraft, but there is very little further information available apart from the nature of the powerplant. This was to be four Heinkel HeS 011 turbojets officially rated at some 2,866lb (1,300kg) st each.

Arado E.555

Projected as a flying-wing jet bomber, the E.555 was envisaged in several versions with various belly-mounted powerplants, including two or three BMW 018, two Jumo 012 or two HeS 011A turbojets. The pilot was to be housed in a semi-reclining position in the nose of the aircraft.

Arado E.555 data

Projekt 6	Powerplant	Three BMW 018 turbojets, 7,496lb (3,400kg) st each
	Maximum speed	632mph (1,015km/hr)
	Range	2,300 miles at 39,370ft (3,700km at 12,000m)
	Wing area	1,722.2ft^2 (160.0m^2)
Projekt 7a	Powerplant	Two BMW 018 turbojets, 7,496lb (3,400kg) st each
	Range	2,672 miles (4,300km)
	Wing area	1,506.9ft^2 (140.0m^2)
Projekt 7u	Powerplant	Three BMW 018 turbojets, 7,496lb (3,400kg) st each
	Maximum speed	621mph (1,000km/hr)
	Range	2,423 miles at 39,370ft (3,900km at 12,000m)
	Wing area	1,722.2ft^2 (160.0m^2)
Projekt 8	Powerplant	Three BMW 018 turbojets, 7,496lb (3,400kg) st each
	Maximum speed	612mph (985km/hr)
	Range	2,423 miles at 39,370ft (3,900km at 12,000m)
	Wing area	1,722.2ft^2 (160.0m^2)
Projekt 10	Powerplant	Three BMW 018 turbojets, 7,496lb (3,400kg) st each
	Maximum speed	618mph (995km/hr)
	Range	2,268 miles at 39,370ft (3,650km at 12,000m)
	Wing area	1,506.9ft^2 (140.0m^2)

Arado E.560

The purpose of this mid-wing monoplane is not clear, but a post-war Royal Aircraft Establishment report considered that it was probably a bomber. Three versions were envisaged, all with four underslung jet units. The first had straight wings with a slight taper which increased in the outer panels, and rounded tips. Some 20° of sweepback was applied to the wings of the second variant, with increased taper on the inner leading edges. The third had wings similar to those of second, but without sweepback.

All three were built in model form and wind tunnel-tested, but development went no further.

Arado E.580

Designing to the RLM specification for a high-speed jet fighter issued on September 8, 1944, the Arado company was competing against four other manufacturers. As it turned out, the winning design, the He 162 *Volksjäger* (see page 72), was remarkably similar in appearance to Arado's E.580.

The Arado design was not accepted and the project remained in a very basic state.

Arado E.580 data	
Role	Single-seat jet fighter
Ultimate status	Design
Powerplant	One BMW 003A turbojet, 1,764lb (800kg) st
Maximum speed	460mph at 19,690ft (750km/hr at 6,000m)
Range	311 miles (500km)
Endurance	54min
Weight	5,512lb (2,500kg) loaded
Span	25ft 4⅔in (7.75m)
Length	26ft 3in (8.00m)
Wing area	107.6ft² (10.0m²)
Armament	One or two MK 108 30mm cannon

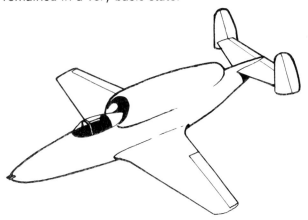

Arado E.581-4

An interesting but somewhat clumsy delta-winged concept, the E.581-4 remained in basic sketch design form only. It had a deep boat-like fuselage which housed a single Heinkel HeS 011A turbojet fed by two side-by-side intakes in the extreme nose. Intended as a single-seat fighter with tricycle undercarriage, it was to be armed with two 30mm cannon. Wing span was 29ft 3½in (8.92m) and length 18ft 5in (5.57m).

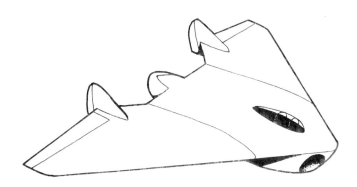

Arado Ar I

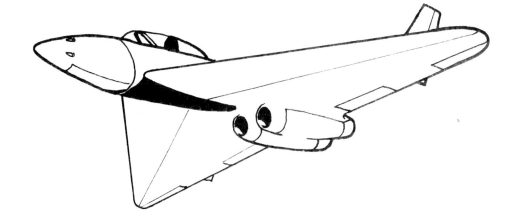

Two Arado designs for two-seat night fighters bore the simple designations Ar I and Ar II (see next entry). The Ar I was to be a clean 35° delta-wing aircraft with twin fins mounted outboard on the wing trailing edges and the crew of two seated in tandem in a pressurised cockpit. Armament consisted of four 30mm revolver cannon housed in the sharply pointed nose, and both versions were to be equipped with centimetric AI radar. A twin-cannon rear turret was mooted and, as the Ar I was also intended to perform the fast bomber role, allowance was made for a bomb load. Powerplant was to be two turbojets, either HeS 011As or BMW 003As, mounted beneath the rear fuselage.

Although the end of the war meant that work on the Ar I and Ar II came to a halt, their numbers bore no "E" prefix and design work seems to have been well advanced.

Arado Ar I data*

Role	Two or three-seat jet night fighter and fast bomber
Ultimate status	Design
Powerplant	Two HeS 011A turbojets (2,866lb st each) or two BMW 003A turbojets (1,760lb st each)
Maximum speed	503mph at 19,670ft (810km/hr at 6,000m)
Endurance	2hr 36min at cruising speed
Weight	27,700lb (12,565kg) loaded, less tail turret.
Span	60ft 4in (18.40m)
Length	42ft 7in (12.96m)
Wing area	710.4ft² (66.0m²) less fuselage section
Armament	Four MG 213 30mm cannon and two 1,102lb (500kg) bombs (plus two 30mm rearward-firing cannon projected)

*With HeS 011A.

Arado Ar II

This sketch design for a 35° swept-wing twin-jet night fighter and fast bomber was, like the Ar I, to be powered by either two HeS 011A or two BMW 003A turbojets. Centimetric AI radar and parachute brakes were to be fitted, together with an armament of four 30mm revolver cannon in the nose. The crew of two or three were to be accommodated in a pressurised cabin fitted with catapult seats. A variety of airframe configurations – including one with a butterfly tail – were studied, but the ending of the war brought all work on the project to a halt.

Arado Ar II data*

Role	Two or three-seat jet night fighter and fast bomber
Ultimate status	Design
Powerplant	Two HeS 011A turbojets (2,866lb, 1,300kg st each) or two BMW 003A turbojets (1,764lb, 800kg st each)
Maximum speed	482mph at 22,970ft (775km/hr at 7,000m)
Service ceiling	37,730ft (11,500m)
Endurance	2hr 12min at cruising speed
Weight	29,057lb (13,180kg) loaded
Span	49ft 2½in (15.00m)
Length	57ft 9in (7.60m)
Wing area	806ft² (75.0m²)
Armament	Four MG 213 30mm cannon and three 1,102lb (500kg) bombs (optional)

*With HeS 011A.

Bachem Ba 349 Natter

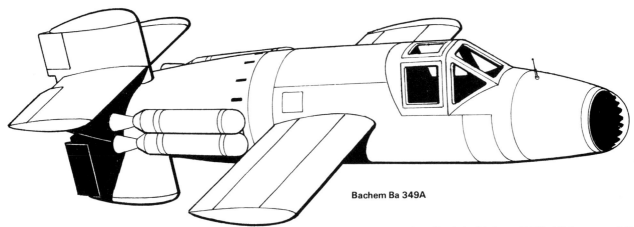

Bachem Ba 349A

As the Second World War drew to a climax it became increasingly important to protect key strategic targets in Germany against Allied bombing, using locally stationed interceptors capable of responding rapidly to attack the bombers within a few miles of their objective. The target-defence interceptors conceived to meet this requirement – which called for a lightweight aircraft with a phenomenal rate of climb, very short take-off run and concentrated offensive ability – represented a radical departure from conventional notions of fighter design. A large number of designs were produced by many aircraft builders, nearly all of them using the short-duration, high-power Walter rocket motor. They were designed to various extremes of sophistication, the Messerschmitt Me 163 being the best known of the more complex types. At the other end of the spectrum was the Bachem Ba 349 *Natter* (Adder). But while the Me 163 took more than six years to reach operational status under *Projekt* X (see DFS 194, page 43), the Ba 349 was not conceived until August 1944. Only eight months after that date a unit of ten Ba 349As was in position at Kirchheim on Teck, ready to tackle USAAF bombers.

Originally given the Bachem-Werke designation BP 20A, the *Natter* was designed to be semi-expendable and very cheap to build. Every possible economy was made, and non-strategic materials were used throughout, except in such critical areas as the motor, fuel tanks and wing mainspar. The single-seat cockpit was located behind a perspex nose cone which protected a battery of 33 R4M 55mm unguided rockets. The wings were short and stubby, and the empennage was of cruciform layout, with twin vertical fins, one above and one below the rear fuselage.

Ba 349A with undercarriage fitted for ground handling purposes (*via Alex Vanags*)

Test pilot Lothar Siebert climbs aboard before the first, unsuccessful, manned test of the Ba 349 (*via Pilot Press*)

The Ba 349 was launched along three near-vertical rails, each 31ft (9.45m) long, and assisted into the air by four jettisonable, 60sec-endurance Schmidding rockets attached to the sides of the fuselage. Until the pilot achieved visual contact with the enemy the aircraft was controlled from the ground via a radio/radar link. At about a mile from the bombers the pilot was to take control and eject the perspex nose cone. A target would then be selected and the rockets launched at it, usually into the bomber's belly. The whole attack had to be completed within about eight minutes of the launch, such was the short duration of the Walter HWK 109-509A rocket motor. On completion of the mission the pilot would eject himself from the cockpit, at the same time jettisoning the rear fuselage, which housed the precious rocket motor. Both pilot and engine would then descend on separate parachutes.

By March 1945 a series of manned test flights had been completed and production had begun. Some 36 Ba 349As had been built by April 1945 but the single unit to become operational never actually saw action, its aircraft being destroyed by Luftwaffe personnel to avoid their being captured by approaching American ground forces.

Manufacturing drawings and specifications for the *Natter* were sold to the Japanese shortly before the war ended in Europe, but only a few partly built examples were found in Japan when the Pacific war ended.

An improved version of the *Natter* was designed in 1945 as the Ba 349B. The single prototype was fitted with a more powerful Walter HWK 109-509C rocket unit with an auxiliary cruising chamber which gave it a 12min endurance and a greater top speed and rate of climb. The empennage also differed, the area of the lower fin being reduced and that of the upper increased.

Bachem Ba 349 data

Role	Semi-expendable single-seat target-defence rocket interceptor
Ultimate status	Operational
Powerplant	One Walter HWK 109-509A rocket motor (3,750lb, 1,700kg) thrust (Ba 349A) One Walter HWK 109-509C or D rocket motor (4,410lb, 2,000kg) thrust (Ba 349B) plus four Schmidding assisted take-off units (1,102lb, 500kg thrust each) or two SG 34 units (2,205lb, 1,000kg thrust each) (both versions)
Maximum speed	560mph at 16,400ft (900km/hr at 5,000m) (Ba 349A), 630mph at 36,090ft (1,105km/hr at 11,000m) (Ba 349B)
Climb rate	29,530ft (9,000m) in 0.80min (Ba 349A), 0.67min (Ba 349B)
Ceiling	52,490ft (16,000m) (both versions)
Range	37 miles at 9,840ft (60km at 3,000m) (Ba 349A)
Weight	4,520lb (2,050kg) (Ba 349A), 5,005lb (2,270kg) (Ba 349B) at take-off
Span	11ft 10in (3.60m) (Ba 349A), 13ft 0½in (4.00m) (Ba 349B)
Length	18ft 9¾in (5.72m) (Ba 349A), 18ft 8¾in (6.02m) (Ba 349B)
Wing area	38.75ft^2 (3.6m^2) (Ba 349A), 50.6ft^2 (4.7m^2) (Ba 349B)
Armament	24 73mm *Föhn* or 33 R4M 55mm unguided rockets (Ba 349A), 24 73mm *Föhn* unguided rockets and two MK 108 30mm cannon (proposed)

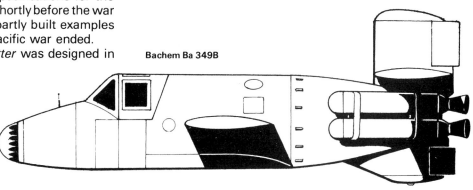

Bachem Ba 349B

Blohm und Voss P.175

Blohm und Voss was among the more prolific aircraft design companies in Germany and produced a long list of jet projects, most of which are poorly documented. One of the earliest jet designs was the P.175, which is known to have been intended as a shipboard fighter powered by a single Junkers turbojet and with a wing span of 20ft 4in (6.20m).

The "P" prefix in these Blohm und Voss designations indicates that the projects were design studies only.

Blohm und Voss P.177

This two-seat fighter-bomber of asymmetric layout was to have been powered by either a single 1,750 hp Jumo 213A piston engine or turbojet engines of unknown number and type. A formidable armament of six MG 151/20 20mm cannon was envisaged, wing span was 39ft 4¼in (12.00m), and all-up weight was 13,133lb (5,950kg).

Blohm und Voss P.178

Another fighter-bomber of asymmetric design, the P.178 was intended to be powered by a single 1,984lb (900kg) st Junkers 004B turbojet and to carry an armament of two MG 151/15 15mm machine guns and a fuselage-mounted bomb of 1,102lb or 2,205lb (500kg or 1,000kg). Wing span was put at 39ft 4¼in (12.00m), and drawings show a pair of tubes projecting from the tail, possibly indicating the use of rockets to boost climb rate during the escape following a dive-bombing attack.

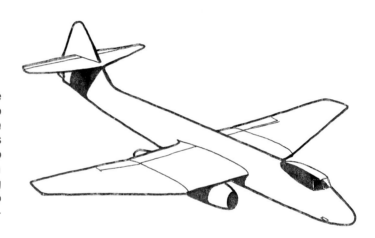

Blohm und Voss P.188

The compound-swept wings of the P.188 long-range bomber project are a good illustration of the problems encountered in designing wings for high-speed jet aircraft. Straight wings have good low-speed and lift characteristics but suffer a high degree of drag. On the other hand, the swept wing has good high-speed qualities but poor low-speed behaviour. Four versions of the P.188 were envisaged, all with the same bomb load and wing shape, the inner section of each wing being swept back at 20° and the outer half swept 20° forward. This had the effect of giving the leading edges a sweep along their whole length, while the "W" planform kept the centre of lift acceptably far forward, theoretically creating better low-speed qualities coupled with good high-speed behaviour. Wing incidence was also to be variable in flight through 12° in order to avoid an excessively nose-high attitude during landing.

The P.188.01 and P.188.03 were to have single fins while the P.188.02 and P.188.04 were to have twin fins and rudders. The four engines were located beneath the wings, separately or in double pods, and all four versions were to land on two tandem-mounted pairs of large fuselage wheels, with single outrigger wheels under each engine pod.

The *Oberkommando der Luftwaffe* (OKL, High Command of the Luftwaffe) issued the specification which resulted in these projects during 1943, seeking a new long-range bomber to replace the Heinkel He 177. A number of jet bomber designs were submitted, the P.188 among them, but the final choice fell on the Junkers Ju 287.

Blohm und Voss P.188.01 data

Role	Long-range two-seat jet bomber
Ultimate status	Design
Powerplant	Four Jumo 004C turbojets, 2,205lb (1,000kg) st each
Maximum speed	512mph at 19,690ft (820km/hr at 6,000m)
Range	1,420 miles (2,285km)
Ceiling	42,650ft (13,000m)
Weight	53,361lb (24,200kg) loaded
Span	88ft 7in (27.00m)
Length	57ft 2in (17.42m)
Wing area	645.83ft² (60.0m²)
Armament	Two MG 151/15 15mm machine guns and 4,410lb (2,000kg) of bombs

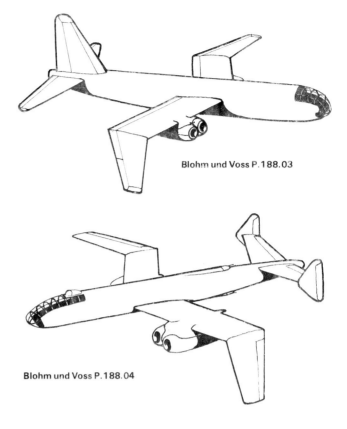

Blohm und Voss P.188.03

Blohm und Voss P.188.04

Blohm und Voss P.190

Details of this projected single-seat jet fighter are limited to the fact that it was to be powered by a single 1,984lb (900kg) st Junkers Jumo 004B turbojet

Blohm und Voss P.194

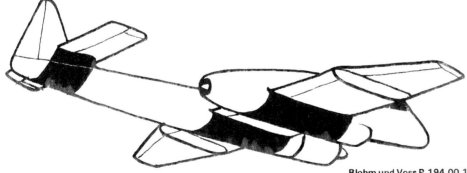

Blohm und Voss P.194.00.110

The P.194 ground-attack and general-purpose project was a development of the single piston-engined BV 141 reconnaissance/observation aircraft. Both aircraft were of asymmetric layout, a result of Dr Ing Richard Vogt's efforts to give the crew good all-round vision.

Twelve examples of the production BV 141 were built and some underwent operational trials on the Eastern Front.

The very much faster P.194, designed in March 1944, was to be powered by a Jumo 004B turbojet as well as the BMW 801D radial piston engine. The main fuselage carried the empennage and radial engine and was offset from the centre of the mainplane. The other unit, similarly offset, carried the pilot and a turbojet. Several other versions were proposed, all basically similar and differing only in small details, but none was built.

Blohm und Voss P.194.01.02 data

Role	Mixed-power multi-purpose aircraft
Ultimate status	Design
Powerplant	One Jumo 004B turbojet (1,984lb, 900kg st) and one BMW 801D radial engine (1,440 hp)
Maximum speed	478mph at 14,760ft (770km/hr at 4,500m)
Range	650 miles (1,045km)
Weight	20,950lb (9,150kg) loaded
Span	50ft (15.25m)
Length	39ft 8¼in (12.10m)
Wing area	391.83ft² (36.4m²)
Armament	Two MK 103 30mm cannon, two MG 151/20 20mm cannon and 2,205lb (1,000kg) of bombs

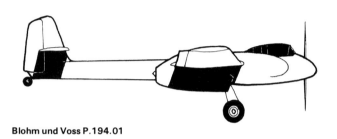

Blohm und Voss P.194.01

Blohm und Voss P.196

Another dive bomber and ground-attack design, the P.196 was to have twin tailbooms and a central fuselage housing the pilot and four cannon in the nose. Beneath the rear of the fuselage were two 1,764lb (800kg) st BMW 003 turbojets in a side-by-side nacelle.

Each of the widely spaced tailbooms extended beneath the mainplane and forward of the leading edges. The nose of each boom was to be used as a bay capable of housing a bomb of up to 1,102lb (500kg). Span was 49ft 2½in (15.00m), length 38ft 4½in (11.70m) and all-up weight about 20,000lb (9,000kg). Development did not proceed beyond the design stage.

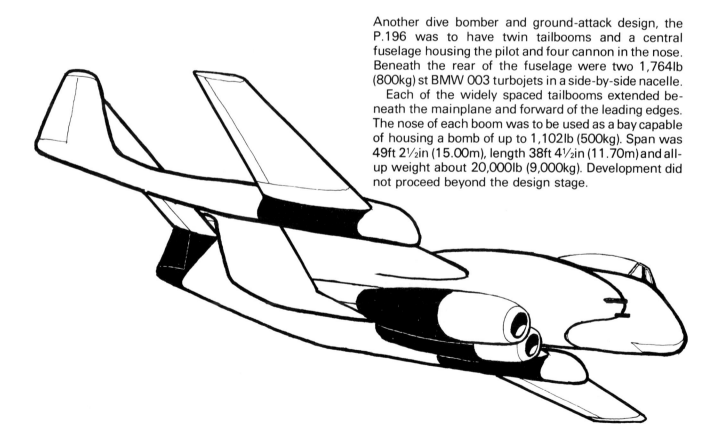

Blohm und Voss P.197

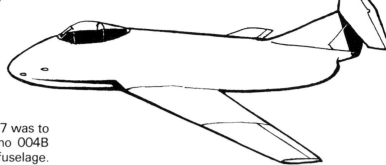

Projected as a single-seat jet fighter, the P.197 was to be powered by two 1,764lb (800kg) st Jumo 004B turbojets mounted side-by-side in the rear fuselage. Wing span was 36ft 5in (11.10m), length 29ft 6in (9.00m), and armament two cannon and two machine guns in the nose. The engine intakes were set just in front of the wing leading edges. A sharply swept fin supported a straight, high-mounted tailplane. Estimated maximum speed at 26,900ft (8,200m) was about 660mph (1,060km/hr) at a weight of 12,853lb (5,830kg). Again, this project went no further than the design stage.

Blohm und Voss P.198

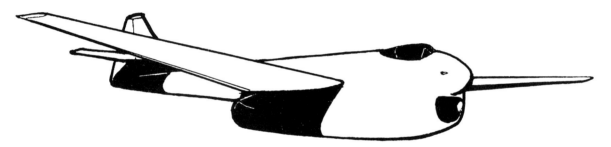

This high-altitude fighter was designed around either a 1,764lb (800kg) st BMW 003 or a 7,496lb (3,400kg) st BMW 018 turbojet fitted in the lower section of the 8ft 2in (2.50m) deep fuselage. The straight, tapering wings had a span of 49ft 2½in (15.00m), sufficient to allow an operating height of around 50,000ft (15,000m). It was estimated that this could be reached in just 11min with the BMW 018 fitted. A swept-wing version was also studied. This project's fate is not known, but it is unlikely to have gone beyond the drawing board.

Blohm und Voss P.198 data*

Role	High-altitude jet fighter
Ultimate status	Design
Powerplant	One BMW 003 turbojet (1,764lb, 800kg st) or one BMW 018 turbojet (7,500lb, 3,400kg st)
Maximum speed	547mph at 11,480ft (880km/hr at 3,500m), 485mph at 49,210ft (780km/hr at 15,000m) (swept-wing version 553mph at 49,210ft, 890km/hr at 15,000m)
Range	900 miles at 50,850ft (1,450km at 15,500m)
Endurance	1hr 50min
Weight	16,000lb (7,440kg) loaded
Span	49ft 2½in (15.00m)
Length	42ft (12.80m)
Armament	One MK 114 55mm cannon and two MG 151/20 20mm cannon

*With BMW 018.

Blohm und Voss P.199

Another high-altitude fighter project, the P.199 is very poorly documented. All that is known is that it was to be powered by a single Junkers Ju 004B turbojet of 1,984lb (900kg) st.

Blohm und Voss P.201

Also projected as a high-altitude fighter, the P.201 is as sparsely documented as the P.199. All that is known is that it was to have been powered by a Walter HWK 109-509A rocket motor.

Blohm und Voss P.202

The principal design feature of this single-seat fighter project was conceived in an effort to overcome the problems of compressibility suffered by straight wings at near-sonic speeds. After take-off the straight wing unit could be swivelled about a central axis through an arc of 35°, sweeping one wing forward and the other

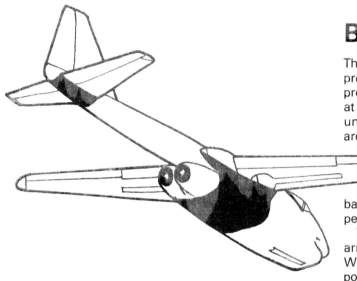

back. This configuration was expected to give good performance at both high and low speeds.

Two 1,764lb (800kg) st BMW 003 turbojets and an armament of three cannon in the nose were to be fitted. Wing span was 39ft 4in (12.00m) in the take-off position and 32ft 10in (10.00m) with the wing fully swivelled.

Blohm und Voss P.203

Outwardly the P.203 looked like a conventional twin-engined aircraft with a BMW 801TJ radial engine on each wing. But closer examination of each elongated engine nacelle revealed a jetpipe exhaust in the extreme rear, with an air intake located midway along the nacelle underside and feeding a turbojet. Each of these units was to be mounted so that it had a centreline that was common with that of the piston engine, or slightly below it. Three versions, with various turbojet engines, were envisaged. The P.203.01 was to have a pair of 2,866lb st HeS 011s, the P.203.02 Jumo turbojets and the P.203.03 a pair of BMW units.

The portion of wing between engine and fuselage was to have greater thickness and chord in order to house the mainwheels of the tricycle undercarriage. The project was not developed.

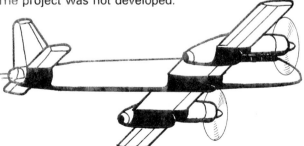

Blohm und Voss P.203 data*

Role	General-purpose and long-range mixed-power fighter
Ultimate status	Design
Powerplant	Two BMW 801TJ radial piston engines (1,800 hp) and two HeS 011A (or Jumo or BMW) turbojets
Maximum speed	416mph (670km/hr) at sea level, 572mph at 39,370ft (920km/hr at 12,000m)
Ceiling	46,910ft (14,300m)
Weight	40,500lb (18,370kg) loaded
Span	65ft 6in (19.95m)
Length	54ft 6in (16.60m)
Armament	Two MG 131 13mm machine guns, two MG 151/15 15mm machine guns, two MK 108 30mm forward-firing cannon, two MG 131 13mm machine guns in remotely controlled rear turret, and 2,205lb (1,000kg) of bombs

*Performance estimates for P.203.01.

Blohm und Voss P.204

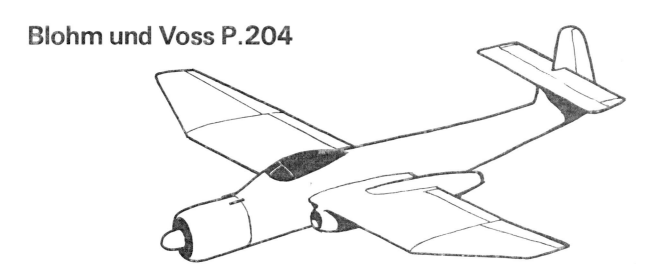

Another mixed-powerplant project, the P.204 was designed for ground attack and dive bombing. The fuselage housed a conventional radial engine in the nose, with a BMW 003A turbojet mounted underneath the port wing and close to the root.

Bombs were to be carried within the fuselage, while a single BV 246 glide bomb could be carried under the fuselage. Other armament consisted of two wing-mounted guns and two propeller-synchronised fuselage guns.

Blohm und Voss P.204 data

Role	Mixed-power single-seat dive bomber and ground-attack aircraft
Ultimate status	Design
Powerplant	One BMW 801D radial engine (1,440hp) and one BMW 003A turbojet (1,764lb, 800kg st)
Maximum speed	404mph (650km/h) at sea level, 472mph at 26,250ft (760km/hr at 8,000m)
Range	330 miles (530km)
Weight	18,730lb (8,500kg) loaded
Span	46ft 11in (14.30m)
Wing area	362.7ft^2 (33.7m^2)

Blohm und Voss P.209

This project was based on two designs, both single-seat fighters but otherwise radically different. The first and better documented version, the P.209.01, had a tailless layout and was powered by a single 2,866lb (1,300kg) st HeS 011A turbojet mounted in the rear of the short fuselage and fed by an air intake in the nose. The wings were swept back, with streamlined pods close to the

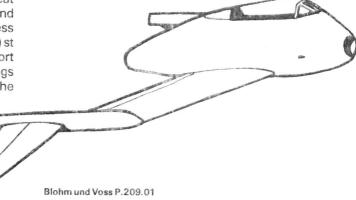

Blohm und Voss P.209.01

tips. Projecting from the rear of the pods were drooping wingtips with trailing-edge control surfaces.

This project seems to have been superseded by the P.210.

Blohm und Voss P.209.01 data

Role	Single-seat jet fighter
Ultimate status	Design
Powerplant	One HeS 011A turbojet, 2,866lb (1,300kg) st
Maximum speed	547mph (880km/hr) at sea level, 615mph at 28,870ft (990km/hr at 8,800m)
Ceiling	36,750ft (11,200m)
Weight	7,650lb (3,470kg)
Span	34ft 11½in (10.65m)
Wing area	140ft^2 (13.0m^2)
Armament	Two MK 108 30mm cannon

The other version of the P.209, the P.209.02, was of more conventional layout, with a normal tailplane. But the wings were swept forward in an effort to overcome the compressibility problems encountered with straight wings at high speed, while at the same time avoiding the low-speed instability problems suffered with swept-back wings. This version was to be powered by a single HeS 011A turbojet fed by a nose intake. Armament was to be three cannon and all-up weight 5,500lb (2,500kg).

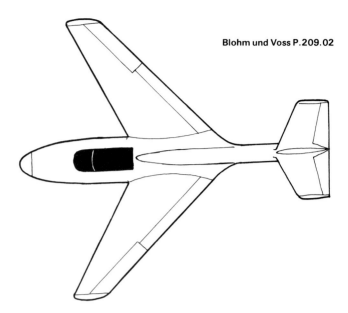

Blohm und Voss P.209.02

Blohm und Voss P.209.02 data

Maximum speed	621mph at 29,530ft (1,000km/hr at 9,000m)
Ceiling	39,370ft (12,000m)
Range	652 miles (1,050km)
Span	26ft 7in (8.10m)
Length	28ft 3in (8.60m)
Wing area	150.7ft^2 (14.0m^2)

Blohm und Voss P.210

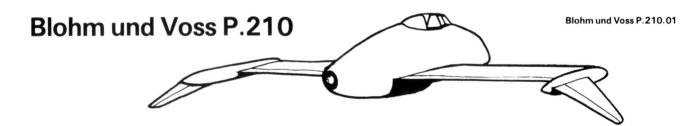

Blohm und Voss P.210.01

Again, this project was envisaged in two different forms. The first was very similar to the P.209.01 in having a tailless layout with wingtip pods and control surfaces, though the P.210's wing span was larger. It differed outwardly from the P.209 in minor details of cockpit canopy location, inwards-retracting main undercarriage, and wingtip pod shape. Powerplant was to be a single 1,764lb (800kg) st BMW 003A turbojet with a nose air intake, and wing span was 37ft 5in (11.40m).

The second P.210 version was of orthodox layout and very simplistic in design, having a straight shoulder wing and empennage. It was to have a centrally mounted HeS 011A turbojet, and span and length were 25ft 8in (7.80m) and 26ft 3in (8.00m) respectively. Both types were intended as fighters but neither was developed.

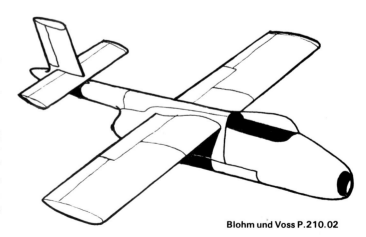

Blohm und Voss P.210.02

Blohm und Voss P.211 Volksjäger

Five companies contested the *Volksjäger* (People's Fighter) competition, the requirements of which were issued on September 8, 1944, by the RLM: Arado, Blohm und Voss, Focke-Wulf, Junkers and Heinkel. Although the Heinkel He 162 ultimately won the contract, the Blohm und Voss entrant, the P.211, was the favourite and probably lost out only because the Heinkel team shrewdly built a full-size mock-up of the He 162 for inspection by the RLM.

The P.211 may also have lost the day because it was too complex to be built quickly and cheaply enough to meet the *Volksjäger* requirement. It was a very much more advanced design than the He 162, having swept wings of riveted steel construction with integral fuel tanks, a delta tailplane and an engine mounted in the fuselage. The fuselage itself was a tubular steel structure to which the cockpit, engine and tailplane were attached. The specification demanded a maximum weight of 4,410lb (2,000kg) and an endurance of at least 30min. Powerplant was to be a single BMW 109-003 turbojet, take-off distance no more than 1,640ft (500m), and armament two 30mm cannon. Finally, and most crucial to the success of the Heinkel design, the *Volksjäger* had to embody a maximum of non-strategic materials and be ready for production by January 1, 1945. Though inferior to the P.211 in many respects, the He 162 met these last two requirements far better.

Blohm und Voss P.211 data

Role	Single-seat jet fighter
Ultimate status	Design
Powerplant	One BMW 003A turbojet, 1,764lb (800kg) st
Maximum speed	472mph (750km/hr) at sea level, 534mph at 26,250ft (860km/hr at 8,000m)
Range	447 miles (720km)
Weight	6,052lb (2,475kg) maximum loaded
Span	27ft 6in (8.40m)
Length	28ft 6½in (8.70m)
Wing area	161.46ft^2 (15.0m^2)
Armament	Two MK 108 30mm cannon

Blohm und Voss P.212

Following the announcement in 1943 that the improved HeS 011 turbojet of 2,866lb st was being developed, a number of manufacturers started designing aircraft around this unit. But it was not until the end of 1944 that an official specification for the Emergency Fighter was issued by the OKL.

This specified the use of the HeS 011A, level speed of 621mph at 22,970ft (1,000km/hr at 7,000m), an operational ceiling of 45,930ft (14,000m), and an armament of four 30mm cannon. The high operational ceiling was demanded because earlier jet engines had

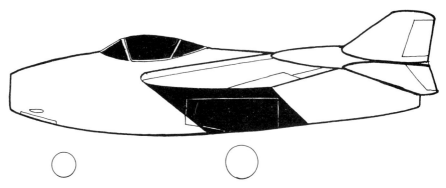

proved reliable only up to 36,000ft (11,000m) which meant that the aircraft they powered could not easily reach the higher-flying Allied aircraft.

Designed to this specification, the P.212 was tailless with a 40°-swept wing. It was to be constructed of metal with a steel skin, and fitted with a pressurised cabin. The Blohm und Voss design was not accepted, however, the Ta 183/1 *Huckebein* (see page 57) being preferred.

Blohm und Voss P.212 data

Role	Single-seat jet fighter
Ultimate status	Design
Powerplant	One Hes 011A turbojet, 2,866lb (1,300kg) st
Maximum speed	565mph (910km/hr) at sea level, 640mph at 29,530ft (1,030km/hr at 9,000m)
Range	699 miles (1,150km)
Service ceiling	42,650ft (13,000m)
Weight	5,954lb (2,700kg) empty, 9,195lb (4,180kg) loaded
Span	31ft 2in (9.50m)
Length	24ft 11in (7.60m)
Wing area	150.7ft² (14.0m²)
Armament	Three MK 108 30mm cannon, or two or three 20mm or 30mm cannon, or 24 R4M 55mm rocket projectiles

Blohm und Voss P.213 Miniaturjäger

During November 1944 the RLM issued a requirement for an extremely simple fighter which could be built even more rapidly than the He 162 *Volksjäger*, though it was not to be semi-expendable like the Ba 349 and was to have a conventional undercarriage. Special attention was to be paid to minimising the complexity of the powerplant, as most other parts could be greatly simplified. It was therefore expected to use a pulsejet to achieve the required performance even though this type of powerplant had already demonstrated bad acoustic characteristics when applied to the Me 328. The production time for a pulsejet was 450 man-hours less than that for a turbojet engine, however, and this settled the matter.

Three companies – Blohm und Voss, Heinkel and Junkers – submitted designs for the *Miniaturjäger*, as the project was known. The Blohm und Voss P.213 was to have all-wood wings and tail and a fuselage made in two halves from pre-formed armoured steel. Take-off, on a tricycle undercarriage, was to be assisted by catapult or rocket.

In the event the pulsejet's power proved inadequate, dropping off steeply with altitude, and the *Miniaturjäger* never left the drawing board.

Blohm und Voss P.213 data

Role	Single-seat pulsejet fighter
Ultimate status	Design
Powerplant	One Argus 109-014 pulsejet, 661lb (300kg) thrust
Maximum speed	435mph (700km/hr) at sea level, 280mph at 29,530ft (450km/hr at 9,000m)
Range	93 miles (150km)
Weight	3,440lb (1,560kg) loaded
Span	19ft 8¼in (6.0m)
Length	20ft 4in (6.20m)
Wing area	53.8ft² (5.0m²)
Armament	One MK 108 30mm cannon

Blohm und Voss P.214

This interceptor project was designed to meet the OKL specification of late 1944. Like several other Blohm und Voss design studies of the time, the P.214 was of tailless layout with wingtip control surfaces (elevons and rudders). The wing was swept back at 40° and had pronounced dihedral. The undercarriage was of the nosewheel type, with the mainwheels retracting forwards into the fuselage. However, the RLM preferred a fighter of more conventional layout and simpler construction, and the P.214 remained just another interesting study.

Blohm und Voss P.214 data

Role	Single-seat jet interceptor
Ultimate status	Design
Powerplant	One HeS 011A turbojet (2,866lb, 1,300kg st)
Maximum speed	603mph (970km/hr) at altitude
Climb rate	4,140ft/min (21m/sec) at sea level
Ceiling	39,450ft (12,400m)
Weight	9,260lb (4,200kg) loaded
Span	31ft 2½in (9.50m)
Length	24ft 3½in (7.40m)
Wing area	150.7ft² (14.0m²)
Armament	Three MK 108 30mm cannon

Blohm und Voss P.215

The P.215 was designed in response to a requirement for a night and all-weather fighter. The unorthodox tailless layout incorporated 30°-swept wings and anhedral wingtip control surfaces. Inboard of these surfaces, at the junction of the wings and wingtips, were small vertical fins. It was to be a large aircraft for its type, with a crew of three in a pressurised cabin, and powered by two side-by-side HeS 011 turbojets housed in the rear fuselage and fed by an air intake in the extreme nose. Also housed in the nose was a very formidable armament of five 30mm cannon, plus radar equipment. Two oblique-mounted upward-firing 30mm cannon and one rear-firing 15mm machine gun were also to be fitted.

Documents refer to another version of the P.215, the P.215.02 with increased fuel capacity and a flying weight of 33,510lb (15,200kg). Full-throttle endurance was increased to 3hr and service ceiling was 39,370ft (12,000m). Neither variant entered development.

Blohm und Voss P.215.01 data

Role	Three-seat night and all-weather jet fighter
Ultimate status	Design
Powerplant	Two HeS 011A turbojets, 2,866lb (1,300kg) st each
Maximum speed	504mph (870km/hr) at sea level, 565mph at 19,690ft (910km/hr at 6,000m)
Service ceiling	41,000ft (12,500m)
Endurance	1hr 20min at 19,690ft (6,000m)
Weight	27,006lb (12,250kg) loaded
Span	61ft 6in (18.75m)
Length	37ft 9in (11.50m)
Armament	Seven MK 108 30mm cannon and one MG 151 15mm machine gun

Blohm und Voss BV 237

Though the BV 237 was originally designed around a single BMW 801D radial piston engine, a later drawing shows a ground-attack version with a supplementary turbojet located centrally beneath the wing. The aircraft's layout was asymmetrical, resembling that of the BV 141, with the main fuselage slightly offset to port on the mainplane and carrying the tailplane and the piston engine. To the starboard of the main fuselage, offset an equal distance from the wing centreline, was a small crew nacelle with one or two seats.

Though the "P" prefix was omitted from this type's designation it seems to have gone no further than the drawing board, and technical information is limited to the piston-engined version.

Blohm und Voss BV 237 data	
Role	Mixed-power dive bomber and ground attack aircraft
Ultimate status	Design
Powerplant	One BMW 801D radial engine (1,440hp) and one turbojet of unknown type
Maximum speed	360mph at 19,690ft (580km/hr at 6,000m) (on piston engine alone)
Range	621 miles (1,000km)
Weight	13,560lb (6,150kg) loaded (minus turbojet)
Armament	Two MG 151/20 20mm cannon, two MK 103 30mm cannon, two rear-firing MG 131 13mm machine guns and six 155lb (70kg) bombs

Blohm und Voss Ae 607

This flying-wing fighter project is poorly documented and no anticipated performance figures are available. However, the proposed powerplant is known to have been a single 2,866lb (1,300kg) st HeS 011 turbojet fed by a nose air intake. The single-seat cockpit was offset to the left of the main fuselage to simplify installation of the centrally mounted engine. The wings were sharply swept back at about 45°, and span was 26ft 3in (7.80m) and length 24ft 3in (7.40m). An armament of three 30mm cannon was envisaged.

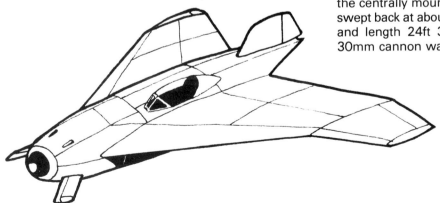

Blohm und Voss Manuell Gesteuertes Raketen Projektil (Manually Controlled Rocket Projectile)

This unusual concept took the form of a man-guided missile which was to be carried to within 180 miles of its target by a Dornier Do 217 bomber. Once within range the missile and the manned command aircraft were to be released before climbing very steeply under rocket power to the stratosphere, where the speed of sound would be attained. Using radar to determine the correct position and speed, the command pilot would then release the rocket projectile and pull up and away, sustaining an astonishing 20g force in the process. This was considered feasible in view of the prone position of the pilot. The missile was then to follow a ballistic trajectory toward its target while the command aircraft returned to land on its central skids.

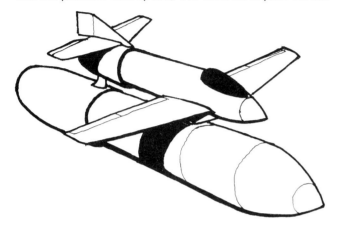

Blohm und Voss Manuell Gesteuertes Raketen Projektil data

Role	Single-seat rocket command aircraft and missile
Ultimate status	Design
Speed	454mph (730km/hr) – Mach 1.0
Range	621 miles (1,000km)
Weight	1,102lb (500kg) command aircraft, 2,640lb (1,200kg) missile unloaded, 8,820lb (4,000kg) total loaded weight
Span	19ft 8in (9.0m)
Length	26ft 3in (8.0m) total, 16ft 3in (4.95m) command aircraft
Wing area	64.58ft² (6.0m²)

von Braun interceptors

Dr Wernher von Braun first submitted plans for a rocket-powered vertical take-off interceptor to the RLM as early as July 6, 1939, following a series of tests with rocket units of his own design. Examples of his oxygen and alcohol liquid-fuelled rockets had been fitted into the extreme rear fuselages of a pair of He 112 single piston-engined fighters and had been subjected to ground testing for some time. The units ultimately each produced a thrust of 2,238lb (1,015kg) for 30sec, and in April 1937 one of the He 112s was flown at this rating at Kummersdorf, only to crash-land after rocket failure. Later in the summer of 1937 a successful rocket flight did take place, and it was found that a 33 per cent increase in speed was achieved at 186mph (300km/hr) when the piston engine was supplemented by the rocket. The von Braun unit was also used to assist He 111 take-offs in 1937. The Walter hydrogen rocket motor was however eventually considered to be more reliable, and the He 112s were consequently refitted with this unit in 1938.

von Braun's first interceptor design was a single-seat aircraft fitted with a newer and much more powerful rocket fuelled with nitric acid and Visol (50/50 mixture of petrol and benzol). An armament of four machine guns, two in each wing root, was to be installed. It was proposed that these aircraft could be stored in a large hangar on pairs of high parallel bars situated about 20ft apart, with the aircraft supported in a vertical position at

von Braun interceptor, first design

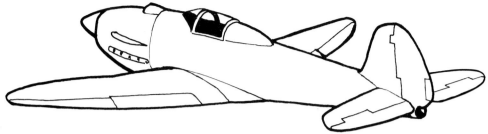

He 112 fitted with Walter rocket motor

a point two-thirds of the way along each wing trailing edge. When needed they were to be run out about 50ft from the hangar on small trolleys running on the bars, taking off vertically from that position.

An extremely powerful rocket motor, rated at 22,400lb (10,160kg) thrust, was expected to propel the aircraft to 25,000ft (7,600m) in 53sec under the control of a gyroscopic guidance system and to permit near-supersonic speeds. On reaching operational height the short-duration take-off rocket would cut out and horizontal flight would be assumed on the power of a cruising chamber rated at 1,700lb (770kg) thrust. The aircraft would be able to fly and climb for a further 15min at a maximum rate of 3,940ft/min (1,200m/min), and after the attack it would glide back and land on a central skid.

The concept was rather too advanced for the time and it was not until the first flight of the Bachem Ba 349 *Natter* in February 1945 that von Braun's ideas were realised.

Following further tests of von Braun's rocket motor a second version was submitted for official approval on May 27, 1941. This development was similar in layout but was to have low rather than mid-set wings, while take-off was to be from the back of a truck which could be sited at almost any location needing defence from bombers. It was intended to perform both the night fighter and interceptor roles but once again failed to win official support.

von Braun interceptor data

Role	Vertical take-off single-seat rocket interceptor
Ultimate status	Design
Powerplant	Nitric acid/Visol rocket motor for take-off (22,400lb, 10,160kg thrust), with cruising chamber (1,700lb, 770kg thrust)
Maximum speed	429mph (690km/hr) on cruising chamber
Weight	11,200lb (5,080kg) loaded
Span	30ft (9.15m)
Length	28ft (8.53m)
Armament	Four machine guns

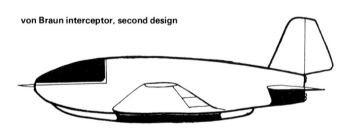

von Braun interceptor, second design

BMW Schnellbomber I

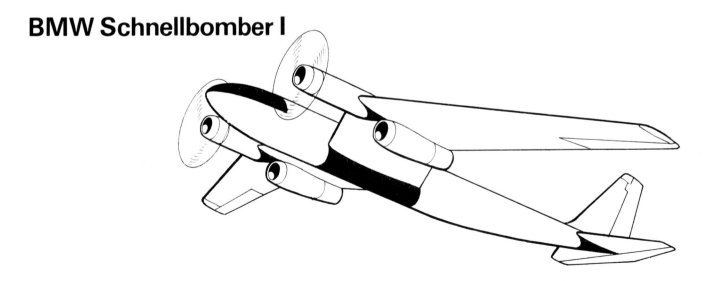

A number of interesting bomber design studies were produced by BMW, all based on the company's own turbojet and turboprop engines. The *Schnellbomber* I (Fast Bomber I) was to be powered by a combination of two BMW 028 turboprops and two BMW 018 turbojets. Each 028 was to drive a pair of four-bladed, 12ft (3.66m) diameter contra-rotating propellers and was mounted in the wing, projecting well forward of the leading edge. Below and to the rear of the turboprops were the 018 turbojets, with their air intakes located in the propeller slipstream.

Compound wing sweep was used, with the sections inboard of the engines being swept well forward and the outer sections swept back. The turboprops were to be the main powerplant, the turbojets being intended for use mainly for take-off and in emergency. The four engines were expected to give a combined static thrust of 26,810lb (12,160kg), equivalent to 35,200hp.

A crew of three were to be housed in a single pressurised cabin in the nose, and the bomb load carried in the large wing roots. The undercarriage was to comprise eight wheels: two pairs of mainwheels under the fuselage, a wheel under each wing, outboard of the engines, and twin nosewheels.

BMW Schnellbomber I data

Role	Three-seat mixed-power heavy bomber
Ultimate status	Design
Powerplant	Two BMW 028 turboprops (6,570eshp each) and two BMW 018 turbojets (7,496lb, 3,400kg st each)
Maximum speed	404mph at 22,970ft (650km/hr at 7,000m) (turboprops only), 503mph at 26,250ft (810km/hr at 8,000m) (all units)
Range	590 miles with 44,090lb load at 34,440ft (950km with 20,000kg at 10,500m), 1,709 miles with 22,046lb load at 37,730ft (2,750km with 10,000kg at 11,500m) (BMW 028s only)
Weight	165,010lb (74,850kg) loaded
Span	146ft 6in (44.68m)
Length	103ft 6in (31.56m)
Wing area	2,690.9ft^2 (250.0m^2)
Armament	Four cannon and 33,000lb (15,000kg) of bombs

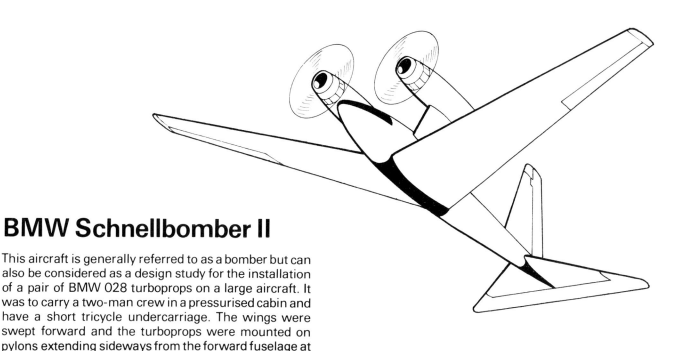

BMW Schnellbomber II

This aircraft is generally referred to as a bomber but can also be considered as a design study for the installation of a pair of BMW 028 turboprops on a large aircraft. It was to carry a two-man crew in a pressurised cabin and have a short tricycle undercarriage. The wings were swept forward and the turboprops were mounted on pylons extending sideways from the forward fuselage at an angle of 45°. Span was to be 116ft 6in (35.53m) and length 70ft 6in (21.50m).

BMW Strahlbomber I

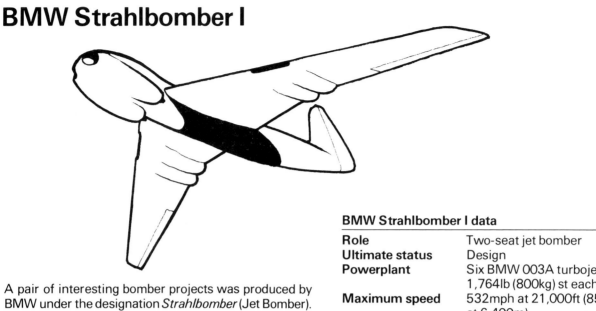

A pair of interesting bomber projects was produced by BMW under the designation *Strahlbomber* (Jet Bomber). Both were tailless and powered by the company's own engines. The *Strahlbomber* I was to be powered by six BMW 003A engines and had an apparently conventional fuselage and fin but no tailplane. The engines were located in pairs, one pair buried in each wing at about one-third span and the third on the nose, with one engine hung on either side of the lower fuselage.

BMW Strahlbomber I data

Role	Two-seat jet bomber
Ultimate status	Design
Powerplant	Six BMW 003A turbojets, 1,764lb (800kg) st each
Maximum speed	532mph at 21,000ft (855km/hr at 6,400m)
Range	1,678 miles (2,700km)
Weight	55,115lb (25,000kg) loaded
Span	85ft (25.90m)
Length	60ft (18.30m)
Wing area	1,076.4ft² (100.0m²)
Armament	8,800lb (4,000kg) of bombs

BMW Strahlbomber II

The second BMW *Strahlbomber* was larger and faster than the first. It was also tailless and was highly unusual in completely lacking vertical control surfaces. Powered by two of the more powerful BMW 018 turbojets, the *Strahlbomber* II was expected to be up to 80mph (130km/hr) faster than the earlier project. The crew of three were to be housed in separate pressurised cabins: the pilot in the nose, the navigator behind him and facing rearwards, and the bomb-aimer lying prone in a blister beneath the central fuselage.

BMW Strahlbomber II data

Role	Three-seat tailless jet bomber
Ultimate status	Design
Powerplant	Two BMW 018 turbojets, 7,496lb (3,400kg) st each
Maximum speed	612mph at 13,120ft (985km/hr at 4,000m)
Range	1,951 miles (3,140km) with 11,023lb (5,000kg) of bombs
Weight	69,400lb (31,480kg) loaded
Span	103ft 6in (31.56m)
Length	60ft (18.30m)
Wing area	1,240ft² (115.2m²)
Armament	Two 20mm cannon in a remotely controlled turret and 11,023lb (5,000kg) of bombs

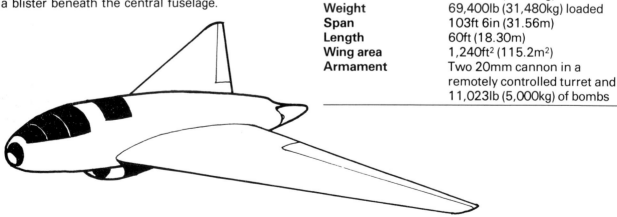

Daimler-Benz Projekt A

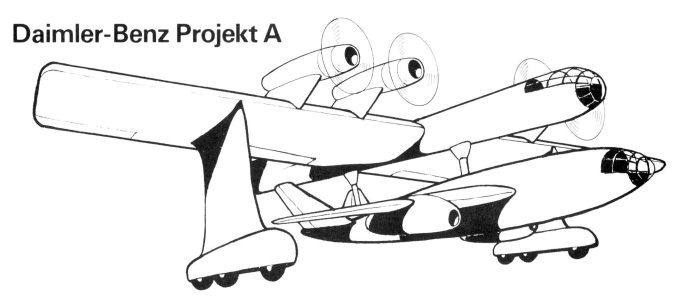

The first of a number of Daimler-Benz projects for turbojet and propeller-turbine aircraft, *Projekt* A was a composite comprising a large "carrier" aircraft with a smaller bomber slung beneath.

The carrier aircraft, designated A I, had a straight mid-wing and four, five or six HeS 021 turboprops, each developing 2,400eshp and mounted on separate over-wing pylons. The undercarriage was to comprise two very long fixed legs with a track of 82ft (25m) and long multi-wheeled bogies. This arrangement was chosen to avoid the possibility of grounding the underslung bomber on take-off, which was to be assisted by jettisonable rocket boosters.

The swept mid-wing A II bomber was powered by two very large turbojets developing 16,555lb (7,500kg) st each and mounted one beneath each wing. It also featured a butterfly tail to allow easier stowage beneath the carrier.

In practice the bomber would have been carried part of the way to the target and then released to complete its mission at high speed while the carrier returned to base.

Daimler-Benz Projekt A data

A I carrier

Powerplant	Four, five or six HeS 021 turboprops, 2,400eshp each
Weight	101,000lb (45,810kg) loaded (four HeS 021s), 114,000lb (51,710kg) loaded (six HeS 021s)
Span	177ft (53.98m)

A II bomber

Powerplant	Two turbojets, 16,535lb (7,500kg) st each
Combat radius	621 miles (1,000km) with full bomb load
Weight	158,290lb (71,800kg) fully loaded
Span	76ft (23.16m)
Armament	66,138lb (30,000kg) of bombs

Daimler-Benz Projekt B

Projekt B was proposed as a carried bomber to be fitted beneath another aircraft designated *Projekt* C. The latter was similar to the *Projekt* A carrier but was designed to be multi-functional and was powered by six 1,900hp DB 603 piston engines, four tractors and two pushers. The composite resulting from the mating of these two designs was known as *Projekt* D.

The bomber was to be powered by a single dorsally mounted turbojet which is quoted puzzlingly as being either a DB 007 developing 2,700lb (1,225kg) st or a massively powerful Daimler-Benz project which was to develop about 28,500lb (12,930kg) st at sea level. As the expected bomb load was 66,138lb (30,000kg) the larger power unit seems more likely, even if its anticipated rating was probably unattainable at that time. Wing span was about 72ft (21.95m), flying weight 154,320lb (70,000kg) and range 621 miles (1,000km) with a full bomb load.

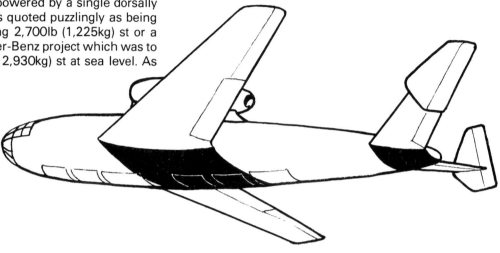

Daimler-Benz Projekt F

Also associated with *Projekt* C was a *Selbst-Vernichtendes* (SV, self-destroying) aircraft or piloted flying bomb coded *Projekt* F. It was proposed that up to six of these aircraft, which bore a strong resemblance to the Fi 103 flying bomb, could be carried beneath *Projekt* C, the combination of *Projekt* F and *Projekt* C being named *Projekt* E.

Compared with the Fi 103, *Projekt* F was much larger and featured swept wings and tailplane. The pilot was located well to the rear, beneath the air intake of the single turbojet. It is not clear what method of attack was to be used, as suicide missions in the Japanese *Kamikaze* style were officially ruled out. The pilot would however have had great difficulty in safely abandoning the aircraft from a position so close to the air intake.

Daimler-Benz Projekt F data

Role	Piloted jet flying bomb
Ultimate status	Design
Powerplant	Single turbojet of unknown type
Maximum speed	665mph (1,070km/hr)
Weight	22,640lb (10,270kg)
Span	29ft 6in (9.0m)
Length	42ft 6in (12.95m)
Warhead	6,614lb (3,000kg) of high explosive

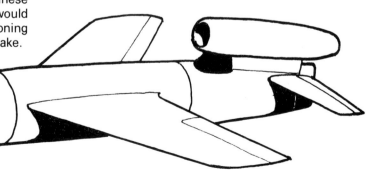

Daimler-Benz P.100/003

Based broadly on the Daimler-Benz composite projects described earlier, this smaller version was to feature a carrier aircraft powered by just four 1,900hp DB 603D piston engines. Consequently the weight of the twin-jet bomber it was to carry was proportionately reduced to match the tug's lower power. The all-up weight of the composite was to be 194,000lb (88,000kg), with a carrier speed of 168mph (270km/hr).

Daimler-Benz P.100/003 bomber data

Role	Composite jet bomber
Ultimate status	Design
Powerplant	Two turbojets of unknown type
Maximum speed	570mph (920km/hr)
Service ceiling	22,970ft (7,000m)
Endurance	6hr 12min
Combat radius	311 miles (500km) with full bomb load
Weight	132,276lb (60,000kg) loaded
Armament	66,138lb (30,000kg) of bombs

DFS 194 (Projekt X)

The DFS rocket aircraft had its origins in two separate research programmes which finally converged when the RLM initiated the highly secret *Projekt* X in 1937. This combined Doctor Alexander Lippisch's tailless aircraft experiments and Doctor Hellmuth Walter's rocket motor work to produce the direct predecessor of the war's most startling combat aircraft, the Me 163 *Komet*.

The Walter liquid-fuel rocket motor programme first got off the ground in February 1937 when a He 72 *Kadett* biplane trainer was used to flight-test one of the earliest units. It ran on hydrogen peroxide with a paste-type catalyst, producing a thrust of 300lb for 45sec. The tests revealed that a great increase in power could be achieved by using a liquid-spray catalyst instead of the paste. By April 1937 the improved Walter rockets had been fitted into a Focke-Wulf FW 56 *Stösser* and a Junkers A 50 Junior for ground testing, producing some 660lb (300kg) thrust for 30sec. Later that year a new all-liquid-fuel motor was refitted into the He 72 for further testing.

DFS 194 in flight during 1940

Eleven years before these experiments, in 1926, Lippisch had built his first tailless glider, named the *Storch*. This was developed ultimately into the *Storch* V in 1929, whereupon Lippisch turned his attention to the delta wing. In 1930 his Delta I glider had been built and flown, and was eventually given a 30hp piston engine.

By 1932 his experiments had attracted the notice of larger manufacturers and his Delta III, with two engines driving fore and aft propellers, was built by Focke-Wulf. Eventually Lippisch's team was taken on by the *Deutsches Forschungsinstitut für Segelflug* (DFS,

The He 72 *Kadett* used to flight-test one of the first Walter liquid-fuel rocket motors

German Research Institute for Gliding). After various setbacks his Delta IVb was given the official RLM designation DFS 39 in 1937.

Powered by a 100hp piston engine, the DFS 39 had a slightly gulled delta wing with pronounced wingtip anhedral. At this point the RLM was looking forward to a second version of the DFS 39 powered by an 88lb (40kg) thrust Walter rocket, representing the beginning of *Projekt* X. The RLM stipulated a speed of 217mph for the DFS 39 under rocket power, but Lippisch had already started design work on the DFS 40 and what was to be the DFS 194. These designs he considered more suitable for rocket power because they had central rudders, so avoiding the possibility of a recurrence of the flutter suffered by the wingtip surfaces of the DFS 39. In the event the rocket-powered DFS 39 did not actually fly, being dropped before completion in favour of the DFS 40.

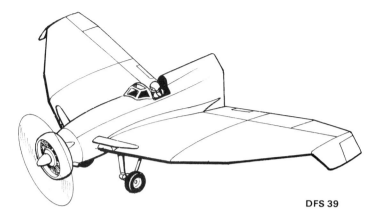
DFS 39

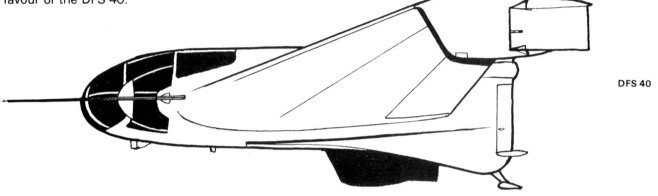
DFS 40

A piston-engined version of the DFS 40, with a 31ft 6in (9.6m) wing span, two seats and delta wings, was built first in 1938. The rocket version was to be built later the same year as a "special medium-speed" testbed with retractable undercarriage. But again, as with the DFS 39, it was dropped before it could be completed, this time in favour of the DFS 194.

When preliminary design work started on the DFS 194 it was once more intended to fit a piston engine at first. But when *Projekt* X and Lippisch's team came under the Messerschmitt umbrella on January 2, 1939, work started on converting it to rocket power. The DFS 194 was completed early in 1939 and ground tests with the Walter HWK R.I-203 "cold" rocket motor began at Peenemünde in October of that year. Four months later the DFS 194 airframe was first test-flown in unpowered glider form. It was first flown under rocket power at Peenemünde-West in August 1940 and subsequently made several highly successful flights, achieving 342mph (550km/hr) in level flight. The DFS 194 was a single-seat tailless aircraft similar in planform to the DFS 40 but lacking the wingtip anhedral, which was compensated for by the fitting of a larger central fin and rudder. The success of the flight-test programme led to the ordering of three high-speed prototypes closely resembling the DFS 194. To be designated Me 163, they turned out to be the first examples of the only reusable rocket-powered interceptor ever to enter service (see Me 163, page 113).

DFS 194 data

Role	Single-seat medium-speed rocket testbed
Ultimate status	Flight test
Powerplant	One Walter HWK R.I-203 rocket motor (661lb, 300kg thrust)
Maximum speed	342mph (550km/hr) at sea level
Climb rate	5,300ft/min (1,615m/min)
Weight	4,620lb (2,095kg) loaded
Span	30ft 6in (9.30m)
Length	17ft 7¾in (5.36m)

DFS 194, 1939 design

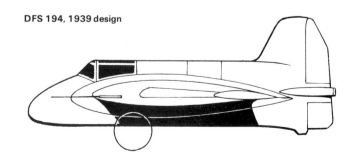

DFS 228

While *Projekt* X was moving towards the eventual production of a high-speed rocket fighter, a related programme started during 1940 was addressing the problems of pressurised cabins, emergency pilot rescue equipment and improved rocket motors for use in the new breed of aircraft. At the centre of this effort was the DFS 228, designed as a rocket-powered flying testbed. By 1941, however, the RLM had changed the specification for the aircraft, requesting that it be made suitable for high-altitude reconnaissance duties. In this role it was to operate at heights of up to 82,000ft (25,000m), unarmed but fitted with special Zeiss cameras. Carried on the back of a Do 217, it was to be released at 32,800ft (10,000m). Height would then be gained and traded off for range in a series of rocket-powered steep climbs and shallow glides, in the course of which a service ceiling of 75,460ft (23,000m) could be attained and a distance of 466 miles (750km) covered. After the reconnaissance mission had been completed a further 185 miles would be covered in the final long glide home from the "safety altitude" of 39,400ft (12,000m). Total range was therefore 650 miles (1,045km) from the moment of release.

The fuselage was constructed in three sections, the forward part being the pilot's pressurised cabin, which was thickly insulated against cold. Temperature and humidity were controlled by a small air-conditioning plant, and if the pressure dropped the pilot could fire four explosive bolts to release the cockpit section from the main fuselage. A single HWK 109-509 rocket unit was housed in the centre section, together with the *C-Stoff* (hydrazine hydrate and methanol) and *T-Stoff* (hydrogen peroxide and oxyquinoline or phosphate) fuel tanks.

The first prototype was completed in 1943 with a Walter HWK 109-509A-1 rocket motor which could be throttled between 661lb and 3,307lb (300-1,500kg) thrust. A maximum speed of nearly 560mph (900km/hr) was expected with this unit. Some 40 flights were made up to a height of 32,800ft (10,000m). The rocket motor was used on none of them and it was later concluded that the cold would have made it difficult to restart for the short bursts required by the operational flight profile.

Nevertheless, the programme went ahead and by the end of the war two DFS 228 prototypes had been flown, and construction work on ten DFS 228A-0 pre-production machines had begun.

DFS 228 data

Role	Rocket-powered single-seat research and high-altitude reconnaissance aircraft
Ultimate status	Flight test
Powerplant	One Walter HWK 109-509A-1 variable-thrust rocket motor, 661-3,307lb (300-1,500kg) thrust
Maximum speed	559mph at 31,820ft (900km/hr at 9,700m)
Range	Up to 650 miles (1,045km)
Service ceiling	75,460ft (23,000m)
Weight	9,284lb (4,210kg) loaded
Span	57ft 7¼in (17.55m)
Length	34ft 8½in (10.56m)
Wing area	323ft² (30.0m²)
Operational equipment	Two Zeiss infra-red cameras

DFS 228 V1 aboard Do 217K carrier aircraft (*via Pilot Press*)

DFS 332

By 1941 the advance of the German rocket research programme had revealed the need for more advanced aerofoil sections and for an aircraft to test them. The DFS 332 flying testbed had a tailless layout, with two side-by-side fuselages. The section of wing to be tested, measuring 14ft 9in (4.5m) across, was to be fitted between the two fuselages. The DFS 332's own lifting surfaces were located outboard of the fuselages, and the angle of incidence of the test specimen could be altered. A crew of two was to be carried, a pilot in the starboard fuselage and the observer in the port. On a typical test flight the DFS 332 was to be towed to a height of around 20,000ft (6,000m+) and then released to attain the desired speed in a glide. Level speed would then be maintained by the use of two Walter R.II-203 rocket motors. Construction was started but the DFS 332 was never actually completed.

DFS 332 data

Role	Rocket-powered research aircraft
Ultimate status	Construction
Powerplant	Two Walter R.II-203 rocket motors, 1,433lb (650kg) thrust each
Normal speed	310mph (510km/hr) in level flight
Weight	7,000lb (3,175kg) loaded
Span	49ft 2¼in (15.0m)
Length	40ft 2in (12.25m)
Wing area	360ft^2 (34.4m^2)

DFS 1068

In the course of its high-speed rocket research programme DFS added to its collection of test aircraft (DFS 194, 228 and 332) a project acquired from Heinkel, the P.1068 (see page 79). Like the DFS 332, it was intended for the testing of wings, and a series of five aircraft were to be built for the study of various wing-sweep angles to determine the most suitable for both high and low-speed flight.

There is some confusion as to the type of engine to be used. An early Royal Aircraft Establishment report gives either BMW 003 or HeS 011 turbojets in varying numbers – which tends to match the original Heinkel P.1068 bomber specification – while later references specify various numbers of Walter rocket motors or even, in the case of the first two examples, no provision for a powerplant at all apart from dummy turbojet nacelles.

According to the latter source, the first two examples, the 1068 and 1068A, were unpowered and were to have 25° and 35°-swept wings respectively. The 1068B had straight wings and rocket propulsion, and the similarly powered 1068C had 35°-swept wings. The final variant, the 1068D, had 35°-swept wings with a laminar-flow profile. The three powered versions were to use up to four Walter 109-509A-1 3,307lb (1,500kg) thrust rocket motors and speeds of around Mach 1 were expected. All five aircraft could be fitted with dummy turbojet nacelles, and all had a common fuselage, the only difference being the centre sections and wings.

The first 1068 was nearly complete when fire destroyed both it and the 50 per cent-complete 1068A and parts of the 1068C.

DFS 346

The final phase of the DFS rocket-powered aircraft research programme saw the embodiment of all the information gathered with the DFS 194, 228, 332 and 1068 in the design of the supersonic DFS 346. The job of actually building the aircraft, working from DFS drawings, was entrusted to the Siebel company in November 1944. Fitted with 45°-swept wings, the DFS 346 was to be carried on top of another aircraft to 32,800ft (10,000m) for its first glide tests and allowed to reach a speed of 560mph (900km/hr). The nose, incorporating a pressure cabin in which the pilot was to lie prone, was connected by explosive bolts to the main fuselage and could be blown off in an emergency to descend by parachute. When low enough the pilot could release himself by means of a catapult and descend on his own parachute. Two Walter rocket motors were to be fitted later in the test programmes, and speed and altitude gradually increased to Mach 2 at

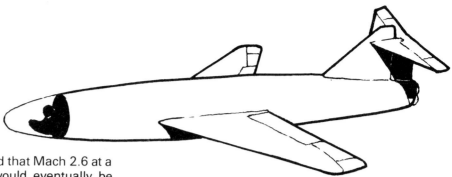

65,620ft (20,000m). It was believed that Mach 2.6 at a height of 114,800ft (35,000m) would eventually be possible.

Although Siebel at Halle had all the necessary materials and the rocket motors, construction work had only just started when the war ended and the town, occupied by US forces, was allocated to the Russians. On Soviet orders the German engineers were set to work on the DFS 346 prototype until October 1946, when the complete plant and all engineers were moved to the Soviet Union. The first prototype was reportedly used for structural tests only, but the second aircraft began glide tests in spring 1948.

For first powered flights the second DFS 346 prototype was carried to 32,800ft (10,000m) by an ex-USAAF B-29 bomber impounded after landing near Vladivostok early in 1945. When released the German pilot accelerated to 683mph (1,100km/hr), which at that height was over Mach 1. In autumn 1951 the former Siebel chief test pilot, Ziese, lost control during a high-speed run from 65,000ft (19,800m) but managed to eject safely; the aircraft was wrecked.

At least another two slightly modified DFS 346 prototypes were completed and flown to gather high-speed flight data in the Soviet Union during the early 1950s.

The original German speed estimates would appear to have been on the optimistic side, as the Russian 346 was reported to suffer some wing flutter at only 683mph (1,100km/hr).

DFS 346 data

Role	Rocket-powered single-seat supersonic research aircraft
Ultimate status	Flight test
Powerplant	Two Walter HKW 109-409B or C rocket motors, 4,410lb (2,000kg) thrust each
Maximum speed	683mph at 32,800ft (1,100km/hr at 10,000m) in Soviet tests. Germans estimated 1,320mph at 65,620ft (2,125km/hr at 20,000m) and 1,715mph at 114,800ft (2,760km/hr at 35,000m)
Span	29ft 2½in (8.90m)
Length	38ft 2¾in (11.65m)
Wing area	215.28ft² (20.0m²)

DOBLHOFF HELICOPTERS

In October 1942 the *Wiener-Neustädter Flugzeugwerke* (WNF) near Vienna started development work on a series of exprimental helicopters designed by Friedrich von Doblhoff. The principal characteristic of these machines was the use of jet-powered rotors. A small fuselage-mounted piston engine was used to drive a single-stage centrifugal compressor, the air from which was mixed with heated petrol vapour before being delivered to the rotor head. From there it passed along each of the hollow aluminium rotor blades to a small combustion chamber at the tip of the blade. There the mixture was ignited, exhausting from the nozzle to drive the rotor.

This novel arrangement had the advantage of eliminating the torque reaction to the fuselage experienced with hub-powered arrangements in which the engine was mounted in the fuselage and geared to the rotor. A further advantage was the reduction of structural weight as a result of the absence of gearing. On the debit side was a very high fuel consumption. The helicopters were however originally envisaged as submarine-borne observation aircraft, and so this drawback appeared to be of little significance.

Some time after the programme began the building of the four experimental helicopters was moved from the Vienna area to Ober-Grafendorf, near St Pölten in lower Austria.

Doblhoff V.I Stepan (WNF 342-1)

The earliest Doblhoff helicopter, named *Stepan*, was flown for the first time in the spring of 1943. It was a single-seater with a 60hp Walter-Mikron piston engine mounted in the main fuselage and powering an Argus As 411 supercharger converted into a compressor. There was no propeller for forward flight, the intention being to operate the type from submarines and small ships in the manner of tethered observation kites.

After a year of testing the V.I was damaged and the opportunity was taken to modify it extensively, bringing it up to V.II standard.

Doblhoff V.I data

Role	Tipjet-powered research helicopter
Ultimate status	Flight test
Rotor diameter	29ft 6in (9.0m) (three-bladed)
Weight	530lb (240kg) empty, 992lb (450kg) loaded

Doblhoff V.II Läufer (WNF 342-2)

In 1944 the much modified V.I re-emerged as the improved V.II, named *Läufer* (Runner). It had been given a more powerful, 90hp, Walter-Mikron engine and a single rudder. The somewhat heavier V.II first flew in 1944, using the same type of blade-tip combustion chambers as the earlier version. But there was still no means of independent forward motion, although it now had directional control by means of the downblast from the rotor being directed on to the rudder. Fuel consumption was still high, in the region of 24 Imp gal/hr (110lit/hr) when hovering.

Doblhoff V.II data

Role	Tipjet-powered research helicopter
Ultimate status	Flight test
Rotor diameter	29ft 6in (9.0m) (three bladed)
Weight	748lb (340kg) empty, 1,014lb (460kg) loaded

Doblhoff V.III (WNF 342-3)

The final two helicopters in the series seem to have been produced in competition with the powered Fa 330, the latter ultimately being preferred as a result of its greater simplicity.

Fitted with a more powerful BMW radial piston engine, the V.III was, for the first time in the series, equipped with a fixed-pitch pusher propeller for forward flight and a small rotor to provide directional control in the hover. One of the final two Doblhoff designs – which of them is not clear – introduced a gyroplaning system designed to cut down fuel consumption. That it might have been the V.III is suggested by a report indicating that the type's engine could be clutched to drive either the Argus 411 compressor supplying the three rotor-tip jets, or the pusher propeller used for forward flight. The resulting fuel saving was said to be significant: 31 Imp gal/hr (140lit/hr) in the hover compared with 9 Imp gal/hr (40lit/hr) in forward flight.

The single-seat V.III was tested until it was eventually wrecked by excessive vibration.

Doblhoff V.III data

Role	Tipjet-powered research helicopter
Ultimate status	Flight test
Powerplant	One BMW Bramo Sh 14A piston engine (140hp) and three rotor-tip combustion chambers (28.8lb/13kg thrust each at 300rpm rotor speed)
Maximum speed	25-31mph (40-50km/hr)
Rotor diameter	32ft 9½in (10.0m)
Weight	1,213lb (550kg) loaded

Doblhoff V.IV (WNF 342-4)

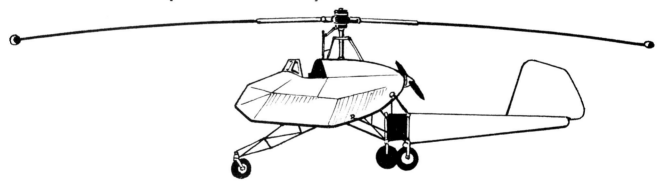

Final design in this series was the V.IV two-seater, which was basically similar to the V.III. A variable-pitch propeller was to be fitted in place of the fixed-pitch unit, eliminating the need for clutch drive and the small yaw-control rotor.

The V.IV was flown in 1945, and when the war ended Friedrich von Doblhoff went to the USA to work for McDonnell as the American company's chief helicopter engineer.

Doblhoff V.IV data

Role	Tipjet-driven research helicopter
Ultimate status	Flight test
Powerplant	One BMW Bramo Sh 14A piston engine (140hp) and three rotor-tip combustion chambers (29.3lb, 13.3kg thrust each at 305 rpm rotor speed)
Maximum speed	31mph (50km/hr)
Rotor diameter	32ft 9½in (10.0m)
Weight	950lb (430kg) empty, 1,410lb (640kg) loaded

Doblhoff WNF 342

The series of four experimental helicopters just described was in fact one of two separate WNF lines of development to bear this designation. A post-war Royal Aircraft Establishment report gives details of a single-rotor helicopter built at the Wiener-Neustadt concern which had a piston engine and rotor-tip jets, and was designated WNF 342. It differed in having tip units larger than those fitted to the V.I-IV. Referred to as "impulse ducts" (pulsejets) rather than combustion chambers, they weighed 9lb (4kg) each, were 36in (90cm) long by 4in (10cm) in diameter, and produced 55lb (25kg) thrust each when operating at 160 cycles/sec.

The single example to be built weighed some 2,646lb (1,200kg) loaded, nearly twice as much as the Doblhoff V.IV (WNF 342-4).

Dornier P.232

The Dornier Do 335 *Pfeil* (Arrow) was one of the most unconventional piston-engined aircraft of the Second World War, yet by the end of the war it was considered important enough to have won a place amongst the four German types earmarked for high-priority development and production status. Bearing the company project designation P.231 and powered by two 1,900hp Daimler-Benz DB 603 radial piston engines, the *Pfeil* was developed to the point of entering operational service over the last two years of the war. But even before the first prototype had flown, a faster version – with a Jumo 004C turbojet replacing the rear DB 603 piston engine – had been designed. This project was designated P.232 and was first mooted in May 1943.

The P.232 was to be a fast bomber and was projected in two forms, one (designated P.3) slightly smaller and faster than the other (P.2).

Dornier P.232 data

Role	Mixed-power single-seat fast bomber
Ultimate status	Design
Powerplant	One Daimler-Benz DB 603G radial piston engine (1,900hp) and one Junkers Jumo 004C turbojet (2,205lb, 1,000kg st)
Maximum speed	503mph (810km/hr) (P.2), 522mph (840km/hr) (P.3)
Weight	12,854lb (5,830kg) (P.2), 12,256lb (5,560kg) (P.3) loaded
Span	45ft 3in (13.80m)
Length	45ft 11in (14.0m) (P.2), 45ft 3in (13.80m) (P.3)
Armament	Two or three cannon and 2,205lb (1,000kg) of bombs

Dornier P.256

This twin-jet fighter was also based on the Do 335, though the fuselage-mounted engines were replaced by a pair of HeS 011A turbojets, one slung beneath each wing. The aircraft's nose was thus left free to house an armament of four 30mm cannon. The crew of two were to be accommodated in separate cabins, the pilot in the normal position and the navigator facing rearwards behind the wing trailing edge.

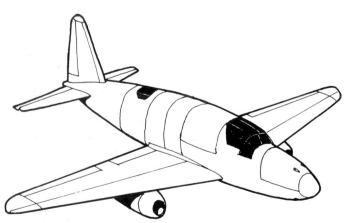

Dornier P.256 data

Role	Two-seat heavy jet fighter
Ultimate status	Design
Powerplant	Two HeS 011A turbojets, 2,866lb (1,300kg) st each
Maximum speed	516 mph at 19,690ft (830km/hr at 6,000m)
Range	621 miles (1,000km)
Weight	27,006lb (12,250kg) loaded
Span	50ft 8in (15.44m)
Length	45ft (13.70m)
Armament	Four MK 108 30mm forward-firing cannon and two MK 108 30mm oblique-firing cannon

Dornier Do 435 (Do 535/He 535)

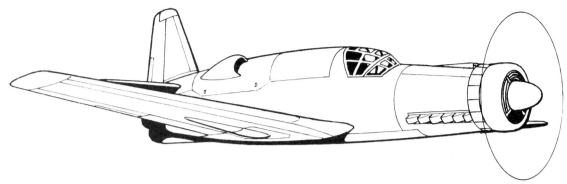

Another version of the Do 335 *Pfeil* was of similar basic layout but with a single Jumo 213J piston engine in the nose and an HeS 011A turbojet in the rear. Other differences from the Do 335 were a wider cockpit to accommodate a crew of two seated side-by-side, a pressurised cabin, radar equipment and modified wings. Designated Do 435, the type was originally proposed as a night fighter powered by Jumo 222 piston engines.

But when the need to find a counter to the RAF's Mosquito night intruders became urgent, Dipl Ing Kurt Tank of Focke-Wulf suggested the use of a mixed powerplant to boost performance.

The mixed-power version has also been referred to under a Heinkel designation, He 535, which could indicate that it was a joint project or one which was transferred from one factory to another.

Dornier canard fighter

No designation has been found for this tail-first (canard) fighter project and little is known about it apart from the fact that it was to be powered by three 2,866lb (1,300kg) thrust HeS 011A turbojets.

Espenlaub E.15

In the late 1920s a number of rocket-powered light aircraft were built and tested as a result of Max Valier's efforts to raise funds for the development of a liquid-fuel rocket motor. Valier was a key member of the *Verein für Raumschiffahrt* (Society for Space Navigation, VfR), and in 1928 he initiated a series of public displays of rocket propulsion to stimulate interest and attract finance. At that time the only rockets that were anywhere near suitable for powering manned vehicles were the little solid-fuel units produced by the pyrotechnist Alexander Sander. These motors developed small amounts of thrust over short periods, but Valier considered them good enough to attract public attention. The money thus raised did not of course go towards solid-fuel rocket research, but instead was spent on developing the more practical but much more complex liquid-fuel motor.

Initial sponsorship was provided by the car builder Fritz von Opel, who engaged Alexander Lippisch and an engineer named Hatry to build a rocket car and aircraft (see Opel-Hatry Rak-1 and Lippisch-Sander *Ente*, pages 133 and 104). Added to the Opel projects was an undertaking by a young aircraft designer named Gottlob Espenlaub, whom Valier and Sander had persuaded to convert to rocket propulsion one of his own EA.1 gliders, which he designed, built and sold.

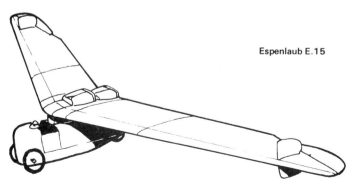

Espenlaub E.15

The glider was renamed the Espenlaub-Valier Rak-3 and originally had two Sander powder rockets mounted one above the other over the centre of the mainplane; this was soon changed to a side-by-side arrangement. A thin sheathing of steel plate was added to protect the fin and upper trailing edge of the wing from exhaust gases. But the unsuitability of the aircraft's layout was soon realised and the experiment was eventually halted after the tail caught fire. Before that, however, rocket thrust had been measured in a series of tethered tests, and the aircraft even flew on a couple of short hops.

While this work was going on Espenlaub had asked a Swiss designer named Sohldenhoff to design a completely new aircraft especially for rocket propulsion. The new aircraft was to be called the Espenlaub E.15. It was of an unusual tailless layout and was designed around a battery of 20lb-thrust Sander cruising rockets mounted above the central wing section. The E.15 was completed during 1929 and first underwent trials powered by a 20hp Daimler piston engine. Once the design's stability had been established the piston engine was removed and the rocket units installed. A number of ground tests were carried out before the E.15 flew for the first time at

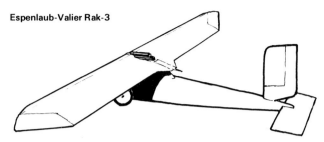

Espenlaub-Valier Rak-3

Bremerhaven on May 4, 1930, in the hands of Espenlaub himself. A number of inconclusive tests followed but the project was promptly brought to an end as far as Espenlaub was concerned when he accidentally fired a 176lb (80kg) thrust take-off rocket motor in mid-flight. The added thrust caused the aircraft to dive into the ground from low altitude. Espenlaub escaped with his life but he dropped the project as a consequence.

By this time the main aim of raising money for development of a liquid-fuelled rocket motor had been achieved. Ironically, the chief instigator of the project, Max Valier, was killed a few days after the E.15 crash when a liquid-fuel rocket motor exploded while he was working on it.

Fieseler Fi 103 Reichenberg

Fieseler Fi 103 R-IV

Development of the Argus pulsejet was started by Dr Ing Gunther Dietrich in 1938 and was given RLM backing in 1939. Work progressed steadily until in April 1941 a Go 145 training biplane was used for flight tests, with an Argus unit fixed between the undercarriage legs. The tests proved satisfactory and an RLM order was placed in June 1942. Six months later the Argus As 014 unit was matched with the Fieseler Fi 103 for development as a flying bomb.

Alternatively known as the FZG-76 or *Vergeltungswaffe* 1 (Reprisal Weapon, V-1), the Fi 103 missile was first used against Britain in mid-June 1944. Carrying a 1,874lb (850kg) high-explosive warhead, they were launched in large numbers against London and Southern England.

While the V-1 was being developed it was realised that the problems of guiding the missile satisfactorily would be great and that the best that could be done in a

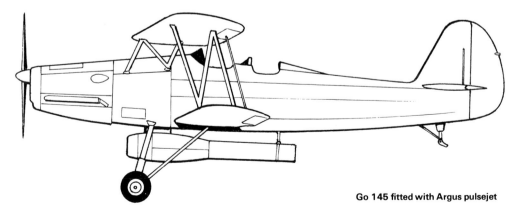

Go 145 fitted with Argus pulsejet

Fieseler Fi 103 R-III (*IWM*)

short time would be a fixed launching ramp to head the missile in the right direction, a gyroscopic system to keep it stable, and a simple form of fuel control. The missile was made to fall to earth in the general target area by the simple expedient of filling its tanks with just enough fuel to take it there; when the fuel ran out, the pulsejet stopped and the V-1 fell on the target.

A big improvement on this arrangement would be a piloted flying bomb, and this was considered at the end of 1943, before the haphazard V-1 campaign against Britain had even started. The concept was soon rejected, though, such was the revulsion against the inevitably suicidal nature of such weapons. But the idea was resurrected in 1944 in the form of the piloted Fieseler Fi 103 V-1, and later in glider form as the Me 328. Work started on the latter in May 1944, but by D-Day the first Me 328 had still to be completed and attention turned to four projected adaptations of the Fi 103. These had been schemed by the DFS and given the cover name *Reichenberg*.

The four variants were converted from standard by the Henschel company in just 14 days. The first, the R-I with a single seat, nose ballast, landing skids and flaps, had no pulsejet and was to be used for basic training. The R-II was also a trainer but had a second seat in the nose for an instructor. The R-III was an advanced trainer similar to the R-I but with an Argus As 014 pulsejet. The final version, the R-IV piloted flying bomb, had a warhead and Argus As 014 pulsejet, and no skids or any other landing aid.

Flight tests, first at Rechlin and then in the hands of the famous woman pilot Hanna Reitsch, began in autumn 1944, when the aircraft was released beneath an airborne He 111. Production of the R-IV was started soon after.

The procedure for the pilot on a mission was to aim the aircraft at the target and then release the cockpit canopy and bale out. But with the engine intake immediately behind him, and the need to abandon the aircraft at the last possible moment, the pilot would seem to have had little chance to escape safely. Some 70 pilots out of thousands of volunteers were selected initially. The declared intention was still to have the pilots bale out and save themselves, but they were nevertheless known as *Selbstopfermänner* ("self-sacrifice men"). The problems of abandoning the aircraft were never really overcome, however, and conflict at official level over what were effectively suicide aircraft meant that *Reichenberg* was never used operationally. Some 175 R-IVs had been completed by the time the programme was halted in February 1945.

Fieseler Fi 103 Reichenberg data

Role	Piloted pulsejet flying bomb
Ultimate status	Flight test
Powerplant	One Argus As 014 pulsejet, 770lb (350kg) thrust
Maximum speed	404mph (650km/hr) in level flight
Weight	4,796lb (2,175kg) (standard V-1)
Span	18ft 9in (5.72m) (R-IV)
Length	26ft 3in (8.00m) (R-IV)
Armament	1,874lb (850kg) high-explosive warhead (standard V-1)

Fieseler Fi 166

Developed from the von Braun concept for a target-defence interceptor, this project was to be powered by a turbojet rather than a rocket motor in the cruise at operational height. Take-off was to be made vertically under the power of a large carrier rocket. This "horse and rider" system called for the release of the take-off rocket once it had burned out, leaving the interceptor to cruise on a smaller integral motor for the remainder of its mission.

FOCKE-WULF Ta 183 PROJECTS

Focke-Wulf P.I

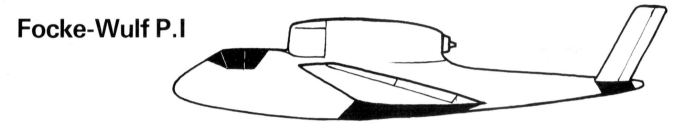

Following an RLM request issued in 1942 for proposals for a single-engined jet fighter, Dipl Ing Kurt Tank designed a number of aircraft under the Focke-Wulf project designation Ta 183 (Ta = Tank).

The first in the series was designed during 1942 and featured 30° forward-swept wings, butterfly V-tail and a single turbojet mounted on top of the fuselage. Three versions were envisaged, the first with a Jumo 004 turbojet, the second with a BMW 003, and a third also with a BMW unit but with straight wings. This project was probably one of the first to be designed exclusively as a single-jet fighter.

Focke-Wulf P.I data

Role	Single-seat jet fighter
Ultimate status	Design
Powerplant	One Jumo 004B turbojet (1,964lb, 900kg st) or one BMW 003A (1,764lb, 800kg st)
Maximum speed	522mph (840km/hr) at sea level, 578mph at 13,120ft (930km/hr at 4,000m)
Weight	6,614lb (3,000kg) loaded
Span	27ft (8.23m)
Length	34ft 4¾in (10.48m)
Armament	Two MK 108 30mm cannon and two MG 151 15mm machine guns

Focke-Wulf P.II

In March 1943 the Focke-Wulf single-jet fighter series was continued with a much less revolutionary design. The straight wings and fin bore a strong resemblance to those of the FW 190 piston engined fighter, and even the orthodox tailwheel undercarriage was retained. A single 1,964lb (800kg) st Jumo 004B turbojet was to be slung beneath the nose, with the pilot sitting above it and forward of the wing leading edge.

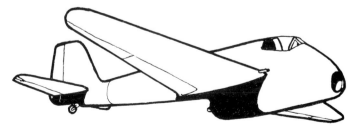

Focke-Wulf P.III

The third project in the Ta 183 series also had an underslung jet engine, this time set further back. The wing leading edges were slightly swept and the trailing edges were straight, while the tailplane was basically similar to that of the previous project. The tailwheel undercarriage was to be replaced with a tricycle arrangement.

Focke-Wulf P.III data

Role	Single-seat jet interceptor
Ultimate status	Design
Powerplant	One Jumo 004B turbojet, 1,964lb (800kg) st
Maximum speed	541mph at 13,120ft (870km/hr at 4,000m)
Range	385 miles (620km)
Ceiling	40,680ft (12,400m)
Weight	7,386lb (3,350kg) loaded
Span	31ft 10¼in (9.70m)
Length	32ft 4in (9.85m)
Wing area	161.46ft² (15.0m²)
Armament	Two MK 108 30mm cannon and two MG 151/20 20mm cannon

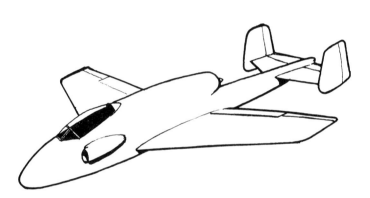

Focke-Wulf P.IV

The fourth Ta 183 design, prepared in November 1943, featured wings with a significant leading-edge sweep and a twin-fin tail unit to avoid interference from the jet exhaust. The single 1,964lb (800kg) st Jumo 004B turbojet was located in the upper fuselage, just behind the cockpit, and exhausted along the whole rear upper length of the tail boom. The air intakes were located on either side of the cockpit, and a tricycle undercarriage was envisaged.

Focke-Wulf P.V Flitzer

The P.V, projected very soon after the P.IV, had a main fuselage and wing very similar to those of its immediate predecessor, and the high-mounted engine was in the same position. To avoid problems with hot gases impinging on the tail boom, the P.V was given twin wing-mounted booms and fins. A high tailplane joined them in the manner of the British de Havilland Vampire.

Two Walter 109-509A liquid-fuel rocket motors were to be located on either side of the turbojet for use as boosters in the climb. Again, the undercarriage was laid out in a tricycle arrangement. Thus far the most radical of Focke-Wulf's single-jet fighter series, the P.V was named *Flitzer* (Madcap).

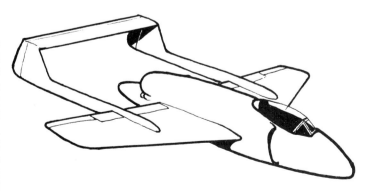

Focke-Wulf P.VI

At the beginning of 1944 Kurt Tank's team produced a design for a swept-wing fighter with a stubby fuselage, sharply swept fin and a high-mounted swept tailplane. It was to have a centrally mounted 2,866lb (1,300kg) st HeS 011A turbojet and, above it, a 2,205lb (1,000kg) thrust liquid-fuel rocket motor offering 3½min of extra power. This project advanced to the mock-up stage.

Although at first the P.VI was not seen as an immediately practicable possibility, a very similar design was eventually adopted as the Ta 183 to meet the OKL's Emergency Fighter specification of February 1945.

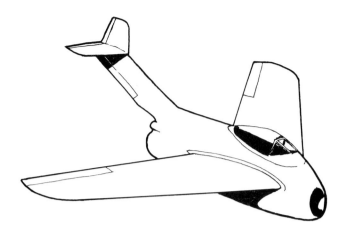

Focke-Wulf P.VII

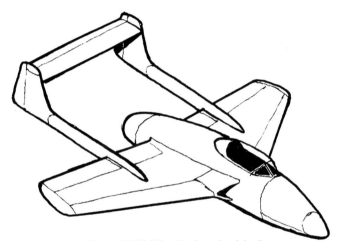

This project was well advanced in development, a full-scale model and some prototype sub-assemblies having been completed, when attention was switched to a swept-wing fighter similar to the P.VI for the OKL's Emergency Fighter competition.

Main version of Focke-Wulf P.VII, with triangular air intakes

The final design in the Ta 183 series resembled the earlier P.V *Flitzer* in having twin booms and a high-set tailplane. The wing leading edge had moderate sweepback and a nearly straight trailing edge. It seems to have been proposed in two outwardly very similar forms. The main version had triangular wing-root air intakes, a slightly drooping pointed nose, and a deeper lower rear fuselage housing a Walter rocket motor to supplement the HeS 011A turbojet. With the rocket operating the P.VII was expected to climb to 37,730ft (11,500m) in under two minutes, while the turbojet-only version was expected to take nearly 25min to reach 36,090ft (11,000m).

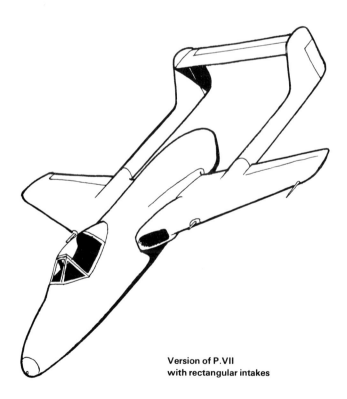

Version of P.VII with rectangular intakes

Focke-Wulf P.VII data

Role	Single-seat mixed-power interceptor
Ultimate status	Construction
Powerplant	One HeS 011A turbojet (2,866lb, 1,300kg st) and one Walter HWK 109-509A rocket motor (3,307lb, 1,500kg thrust)
Maximum speed	516mph at 19,690ft (830km/hr at 6,000m)
Range	800 miles (1,290km)
Weight	10,472lb (4,750kg) loaded
Span	26ft 3¼in (8.0m)
Length	32ft 1¼in (9.78m)
Armament	Two MK 108 cannon and two MG 151/20 20mm cannon

Wooden mock-up of secondary version of P.VII (via *Pilot Press*)

Focke-Wulf P.VIII

Though this project is included in the Ta 183 series, it did not in fact meet the "single jet" requirement in that it was to be powered by an HeS 021 turboprop. The layout was basically similar to that of the P.VII, even down to the engine location. But instead of exhausting through a tail jetpipe the 3,300eshp engine drove a nose-mounted propeller via a long shaft running forward beneath the cockpit.

Focke-Wulf Ta 183 Huckebein

Having been shelved in favour of the twin-boom P.VII, the P.VI was resurrected later in 1944 and slightly modified to meet a new specification, that for the OKL's Emergency Fighter. Two variants of the *Huckebein* (Raven), as it was named, were submitted in competition with six designs from four other manufacturers. The specification called for a fighter powered by the HeS 109-011A turbojet of 2,866lb st, which had been under development since 1943. The new fighter was to have a maximum speed of 621mph at 22,970ft (1,000km/hr at 7,000m) and the ability to operate at up to 45,930ft (14,000m) in order to reach the Allies' high-flying B-29 bombers and Mosquitos.

The Ta 183/I was little different from the original P.VI design, except that the tail-mounted rocket was removed and the wing had greater chord. Comprising 40°-swept wings and a 60°-swept fin, the flight surfaces were made mainly of wood with Dural main spars. The armament of four 30mm cannon was arranged around the nose air intake.

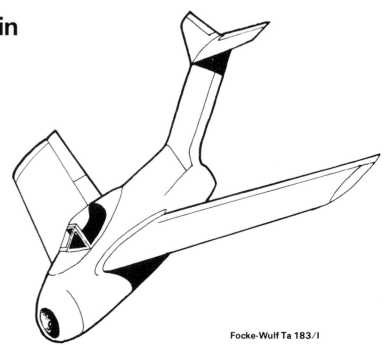

Focke-Wulf Ta 183/I

Wind-tunnel model of Ta 183 (*via Pilot Press*)

This variant was the eventual winner of the competition and was accepted in February 1945, 16 prototypes being ordered a month later. The first three prototypes were to have Jumo 004B turbojets, the 11 pre-production prototypes were to be fitted with HeS 011As, and the last two were earmarked as static testbeds. In the event none were actually built, the factories having been captured by the Allies in April 1945.

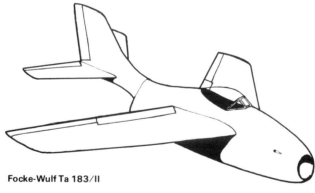

Focke-Wulf Ta 183/II

A second design was tendered in parallel with the winning entry. This was the Ta 183/II, of similar layout but with the cockpit set further back and wing sweep reduced to 35°. The tail was still swept but of a more conventional layout, with the tailplane at the base of the fin. Although slightly faster the Ta 183/II was not chosen for production, probably because it would have taken longer to produce than the Ta 183/I, which is known to have been designed with ease of construction in mind.

Focke-Wulf Ta 183/I data

Role	Single-seat jet interceptor
Ultimate status	Design
Powerplant	One HeS 011A turbojet, 2,866lb (1,300kg) st (provision could also be made for a supplementary rocket motor)
Maximum speed	541mph (870km/hr) at sea level, 590mph at 22,970ft (950km/hr at 7,000m)
Range	615 miles (990km) (full thrust)
Ceiling	45,930ft (14,000m)
Weight	6,240lb (2,830kg) empty, 11,200lb (5,080kg) maximum loaded
Span	32ft 9¾in (10.0m)
Length	30ft 10¼in (9.40m)
Wing area	242.19ft^2 (22.5m^2)
Armament	Two or four MK 108 30mm cannon and 1,102lb (500kg) of bombs

Focke-Wulf Ta 183/II data

Role	Single-seat jet interceptor
Ultimate status	Design
Maximum speed	562mph (905km/hr) at sea level, 597mph at 22,970ft (960km/hr at 7,000m)
Ceiling	44,290ft (13,500m)
Weight	5,844lb (2,650kg) empty, 9,150lb (4,100kg) loaded
Span	31ft 2in (9.50m)
Length	28ft 8½in (8.74m)
Wing area	245.4ft^2 (22.8m^2)

Focke-Wulf P.188

This single-seat fighter project is very poorly documented. All that can be deduced is that it was to be jet-powered, since it had an estimated maximum speed of 662mph (1,065km/hr).

Focke-Wulf FW 190 turbojet variant

One of the earliest Focke-Wulf jet-powered projects was this 1941 design, which consisted simply of a basic FW 190 single-seat fighter airframe with a Focke-Wulf-designed turbojet (rated at 1,323lb, 600kg thrust) in place of the BMW 801 piston engine.

The engine comprised a two-stage radial compressor, single-stage turbine and an annular combustion chamber. There was no jetpipe as such, the exhaust passing through an annular outlet running completely around the circumference of the fuselage immediately in front of the wing leading edge. The result looked like an FW 190 with an elongated engine cowling and no propeller. Development was discontinued in 1942.

The internal fuel tanks had a total capacity of 310 Imp gal (1,410lit).

Focke-Wulf FW 190 turbojet variant data

Role	Single-seat jet fighter
Ultimate status	Design
Powerplant	One FW T.I turbojet, 1,323lb (600kg) st
Maximum speed	472mph (760km/hr) at sea level, 516mph at 29,530ft (830km/hr at 9,000m)
Endurance	1hr 12min
Climb rate	5,412ft/min (1,650m/min) at sea level
Weight	8,267lb (3,750kg) loaded
Span	34ft 5½in (10.50m)
Length	29ft (8.84m)
Wing area	Wing area 196.98ft² (18.3m²)
Armament	Two MG 17 7.9mm machine guns and two MG 151 15mm machine guns

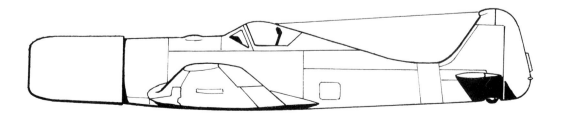

Focke-Wulf Triebflügel

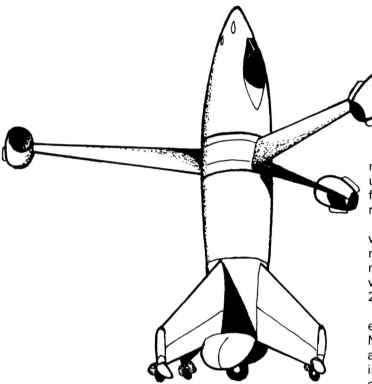

This extremely interesting design was conceived as a vertical take-off target-defence interceptor. Projected by Heinz von Halem in September 1944, it was designed to take off from almost any patch of level ground, using an undercarriage of five tailwheels, one mainwheel in the rear fuselage and a stabiliser in each of the four fins. The proposed method of take-off remains unique to this day. The three untapered variable-incidence wings, fixed to a rotary collar located about one-third of the way down the fuselage, were to rotate around the fuselage, each powered by a Pabst ramjet at the wingtip. The gigantic propeller thus formed would lift the aircraft vertically, incidentally transmitting little or no torque to the fuselage.

Ramjets do not begin to function until they are travelling through the air at some speed, usually between 150 and 200mph, and so the wings had to be accelerated up to operating speed by means of a Walter rocket motor incorporated into each of the three ramjets. With the rotary wings in neutral pitch, the rockets were to power them up to a speed at which the ramjets could be started. The wings would then be moved into fine pitch so that they could generate lift; this was supplemented by the downward component of the thrust from the ramjets.

In the climb ramjet speed was expected to reach a maximum of 670ft/sec (205m/sec) at a time when the aircraft's own airspeed was much lower. This arrangement thus overcame one of the main objections to the use of ramjets, in that the powerplant could continue to function even when the aircraft was flying slower than ramjet operating speed.

On reaching operational height and levelling out, wing pitch could be coarsened to give a rotational speed much closer to the forward speed of the aircraft, so maintaining a constant Mach number of 0.9 at the wingtips. At this point the wings would be rotating at 220rpm.

While some technical problems might well have been encountered, the system offered several advantages. No runway was needed, the powerplant was simple, any fairly volatile fuel could be used, and performance included a high ceiling, low fuel consumption and high efficiency.

Construction was not started, but a wind-tunnel model was tested up to a speed of Mach 0.9.

Focke-Wulf Triebflügel data

Role	VTOL rotary-wing target-defence interceptor
Ultimate status	Design, wind-tunnel tests
Powerplant	Three 2ft 3in (0.68m) diameter Pabst ramjets (1,840lb, 840kg thrust each) and three Walter starter rockets (661lb, 300kg thrust each)
Maximum speed	621mph (1,000km/hr) at sea level, 522mph at 45,930ft (840km/hr at 14,000m)
Range	404 miles (650km) at sea level and 575mph (925km/hr), 1,500 miles (2,410km) at 45,930ft (14,000m)
Endurance	42min at sea level, 3hr 24min at 45,930ft (14,000m)
Climb rate	8.2 sec to 3,280ft (1,000m), 11.5min to 49,210ft (15,000m)
Weight	7,056lb (3,200kg) empty, 11,410lb (5,175kg) loaded
Rotor diameter	35ft 5in (10.80m)
Length	30ft (9.15m)
Armament	Two MK 103 30mm cannon and two MG 151/20 20mm cannon

Focke-Wulf Ta 283

Focke-Wulf Ta 283

While the ramjet has some very obvious drawbacks, it is also a very simple form of powerplant. It was thus increasingly considered as a serious proposition during the latter part of the war, as Germany's supplies of strategic materials dwindled and her need for high-performance fighters became ever more urgent. One of the principal ramjet applications was as a booster for piston-engined aircraft.

Do 17 flight-testing Pabst ramjet

In 1942 ramjet development was accelerated to the point at which a large Pabst unit, mounted on the back of a Dornier Do 17Z, could be tested. The tests were in part preparation for the production of an all-ramjet-powered fighter being designed by Focke-Wulf. The experimental engine was expected to produce anything from 2,400ehp to 10,850ehp, but the tests were abandoned at an early stage because one of the main applications for the ramjet, the Ta 283, failed to materialise. Ramjet experiments did not end there, however, and flight tests were carried out towards the end of the war with an even larger, 20ft-long, Sänger ramjet fitted to the back of a Dornier Do 217E-2. This unit developed some 20,000ehp.

Though the Ta 283 was not built, plans for this little single-seat fighter were well advanced at the end of the

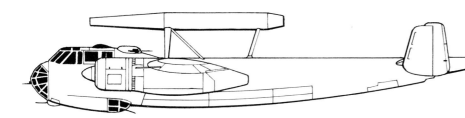

Do 217 with Sänger ramjet

war. It had a long, pointed nose, the wings were swept back at an angle of 45°, and the ramjets were mounted at the extremities of the sharply swept tailplane surfaces. A Walter rocket motor located in the rear fuselage was to be used for take-off and acceleration up to ramjet operational speed.

The ramjets were large affairs, measuring 8ft 9½in (2.68m) long by 4ft 4¾in (1.35m) in diameter, and were expected to develop 2,270hp at 36,090ft (11,000m) and 10,850hp at sea level. Two fuel tanks were to hold a total of 520 Imp gal (2,365lit) for the ramjets, giving the aircraft an endurance of just 13min at sea level or 43min at 36,090ft (11,000m).

A take-off thrust of 6,614lb (3,000kg) for 33sec was to be provided by the Walter starting rocket. This left enough fuel for acceleration and climb-out in case of an unsatisfactory first landing approach. Rocket fuel totalled 89 Imp gal (405lit) of hydrogen peroxide and 36 Imp gal (165lit) of hydrazine hydrate.

Focke-Wulf Ta 283 data

Role	Single-seat rocket/ramjet fighter
Ultimate status	Design
Powerplant	Two Pabst ramjets (10,850ehp each) and one Walter rocket (6,614lb, 3,000kg thrust)
Maximum speed	684mph (1,100km/hr) at sea level, 590mph at 36,090ft (950km/hr at 11,000m)
Range	143 miles (230km) at sea level, 435 miles at 36,090ft (700km at 11,000m)
Weight	11,880lb (5,390kg) loaded
Span	26ft 1in (7.95m)
Length	37ft 8¾in (11.5m)
Wing area	204.5ft² (19.0m²)
Armament	Two MK 103 30mm cannon

Focke-Wulf Ta 400

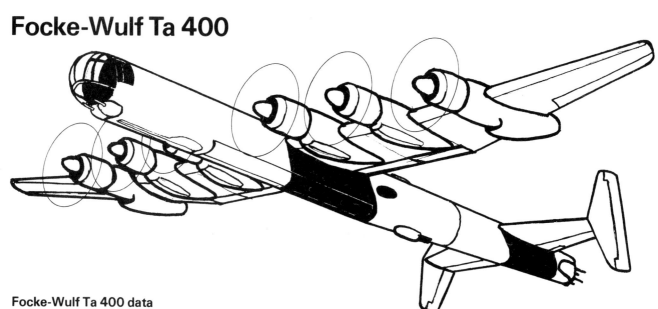

Focke-Wulf Ta 400 data

Role	Mixed-power long-range reconnaissance aircraft and heavy bomber
Ultimate status	Design
Powerplant	Six BMW 801D piston engines (1,700hp each) and two Junkers Jumo 004B turbojets (1,984lb, 900kg st)
Maximum speed	323mph (520km/hr) (piston engines alone), 453mph (730km/hr) (plus turbojets at 16,400ft, 5,000m)
Range	3,107 miles (5,000km)
Weight	132,276lb (60,000kg) loaded
Span	137ft 8½in (42.0m)
Length	92ft 4¼in (28.15m)
Wing area	2,029ft² (188.5m²)
Armament	Ten MG 151/20 20mm cannon in five remotely controlled twin barbettes, and 22,046lb (10,000kg) of bombs

This long-range reconnaissance bomber was a development of the FW 300 and initially was to be powered by six 1,700hp BMW 801D piston engines; a later version also had two Jumo 004 turbojets. Design work on the mixed-power version started early in 1943, the piston engines being mounted conventionally, three to a wing, and the turbojets located one in each of the outer piston-engine nacelles. A crew of six was to be carried, and they shared the task of operating five remote-control twin 20mm cannon barbettes.

The Ta 400 was designed to carry a bomb load of 22,046lb (10,000kg) in a 30ft (9.15m) long bomb bay over a range of 2,980 miles (4,800km). Although the project was of Focke-Wulf origin, several companies in France, Italy and Germany were commissioned to design and build many of the main components. The main spars were to be of dural and the skin of dural sheet. Plans for construction of the Ta 400 were well advanced when the end of the war halted the project.

Focke-Wulf 1000 × 1000 × 1000 bomber

During 1944 three design studies were performed by Focke-Wulf in response to a specification which called for a medium jet bomber capable of carrying a bomb load of 1,000kg (2,205lb) over a range of 1,000km (621 miles) at 1,000km/hr (621mph). No defensive armament was foreseen, the aircraft being considered fast enough to evade interception.

First in the series, *Projekt* A, was a mid-wing aircraft of conventional layout, with the wings swept back at 35° and swept tailplane and fin. The pilot, the only crew member, was located in a cockpit near the nose.

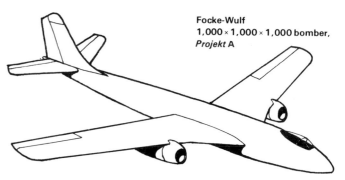

Focke-Wulf 1,000 × 1,000 × 1,000 bomber, *Projekt* A

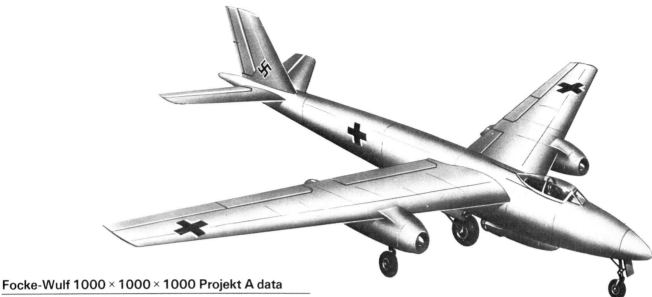

Artist's impression of *Projekt* A in Luftwaffe markings (*Gert W. Heumann*)

Focke-Wulf 1000 × 1000 × 1000 Projekt A data

Role	Medium jet bomber
Ultimate status	Design
Powerplant	Two HeS 011A turbojets, 2,866lb (1,300kg) st each
Maximum speed	621mph at 32,800ft (1,000km/hr at 10,000m)
Range	1,500 miles (2,415km)
Weight	17,860lb (8,100kg) loaded
Span	41ft 6in (12.65m)
Length	46ft 6in (14.17m)
Wing area	290.6ft² (27.0m²)
Armament	2,205lb (1,000kg) of bombs

Drawings show an aircraft with two HeS 011A turbojets hung one beneath each wing, and a wing span of a little over 40ft (12.2m). The specification was expected to have been met on two counts and exceeded by 50 per cent on the third, range.

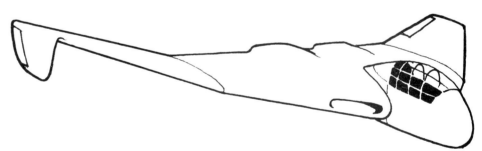

Focke-Wulf 1,000 × 1,000 × 1,000 bomber, *Projekt* B

Projekt B was of an altogether different, flying wing, layout. The wing leading edges were swept back at 45° and the air intakes for the two HeS 011 turbojets were accommodated in the wing roots, on either side of the projecting cockpit and nose section. The centre-section trailing edge was straight and housed the jetpipes. The trailing edges of the outer sections were swept at about 25° and at their tips carried the vertical control surfaces, which projected downwards.

Focke-Wulf 1000 × 1000 × 1000 Projekt B data

Role	Tailless medium jet bomber
Ultimate status	Design
Powerplant	Two HeS 011A turbojets, 2,866lb (1,300kg) st each
Maximum speed	621mph at 32,800ft (1,000km/hr at 10,000m)
Weight	17,860lb (8,100kg) loaded
Span	45ft 11in (14.0m)
Wing area	592ft² (55.0m²)
Length	32ft (9.75m)

Focke-Wulf 1,000 × 1,000 × 1,000 bomber, *Projekt* C

Projekt C, the final design in the series, bore a strong resemblance to the first, but with swept wings positioned slightly higher in the shoulder position. Two HeS 011A turbojets were again the main powerplant, hung one beneath each wing and set at an angle to the chord line in an effort to minimise asymmetric effects in the event of an engine failure.

Focke-Wulf 1000 × 1000 × 1000 Projekt C data

Role	Medium jet bomber
Ultimate status	Design
Powerplant	Two HeS 011A turbojets, 2,866lb (1,300kg) st each
Maximum speed	630mph at 32,800ft (1,015km/hr at 10,000m)
Weight	17,860lb (8,100kg) loaded
Span	41ft 6in (14.17m)
Length	46ft 6in (12.65m)
Wing area	290.6ft² (27.0m²)

Focke-Wulf Volksjäger

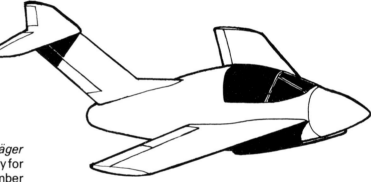

This little rocket fighter was given the name *Volksjäger* but it is not clear whether it was intended as an entry for the RLM People's Fighter competition of September 1944, as it did not properly meet that specification. Its endurance and powerplant, a few minutes and a Walter HWK 109-509 rocket motor respectively, fall short of the RLM demand for 30min and a BMW 109-003 turbojet. It had a mid-wing layout, with 40°-swept wings and high-set swept tailplane. The fuselage was pointed and stubby, with a single central landing skid. Basically similar to the Ta 183 *Huckebein*, it was smaller and had a solid nosecone.

Focke-Wulf Volksjäger data

Role	Single-seat rocket fighter
Ultimate status	Design
Powerplant	One Walter HWK 109-509A-2 rocket motor, 3,750lb (1,700kg) thrust
Weight	4,700lb (2,130kg) loaded
Armament	Two MK 108 30mm cannon

Focke-Wulf FW 272

A number of Focke-Wulf multi-purpose fighter projects were designed around a combination of a single Junkers Jumo 222 piston engine and two BMW 003A turbojets. The FW 272 design lacked the "P" (*Projekt*) prefix in its company designation, suggesting that it may have been the final version of the series. It was to be an all-weather fighter with 30°-swept wings, a maximum speed of 547mph (880km/hr) and a service ceiling of 45,930ft (14,000m).

Focke-Wulf P.0310251-13

This night and bad-weather fighter was to be powered by the Jumo 222/BMW 003 combination which characterised this series. Powerplant arrangement was unusual, the piston engine being housed in the fuselage and driving a rear pusher propeller, and the turbojets being hung one beneath each of the moderately swept wings. Basic layout of the aircraft was mid-wing with a cruciform tail similar to that of the Do 335.

Focke-Wulf P.0310251-13 data

Role	Mixed-power night and bad-weather fighter
Ultimate status	Design
Powerplant	One Jumo 222 piston engine (2,500 hp) and two BMW 003A turbojets (1,764lb, 800kg st each)
Maximum speed	547mph at 31,170ft (880km/hr at 9,500m)
Service ceiling	45,860ft (14,100m)
Endurance	5hr (without using wing jets)
Weight	26,456lb (12,000kg)
Span	68ft 10½in (21.0m)
Length	54ft 3¾in (16.55m)
Wing area	592ft² (55.0m²)
Armament	Four MK 108 30mm cannon or two MK 103 30mm cannon and two MG 213 30mm revolver cannon

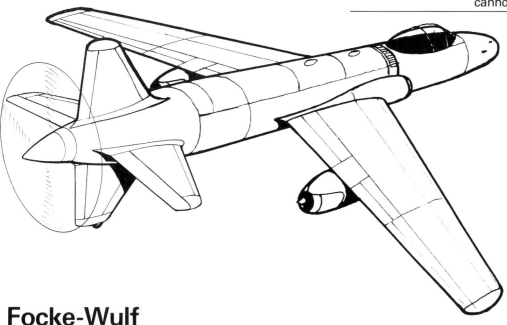

Focke-Wulf JP.000.222.004

Next in the Jumo 222/BMW 003-powered series of heavy fighters was the JP.000.222.004, intended for use as a three-seat day and night fighter.

Focke-Wulf JP.000.222.004 data

Role	Mixed-power three-seat day and night fighter
Ultimate status	Design
Powerplant	One Jumo 222 piston engine (2,500 hp) and two BMW 003A turbojets (1,764lb, 800kg st each)
Maximum speed	547mph at 37,070ft (880km/hr at 11,300m)
Weight	26,500lb (12,020kg) loaded
Span	68ft 10in (20.98m)
Length	47ft 10½in (14.60m)
Armament	Four cannon

Focke-Wulf JP.000.222.010

Developed from the preceding Focke-Wulf project, the JP.000.222.010 was to be used as an all-weather fighter.

Focke-Wulf JP.000.222.010 data

Role	Mixed-power all-weather fighter
Ultimate status	Design
Powerplant	One Jumo 222 piston engine (2,500 hp) and two BMW 003A turbojets (1,764lb, 800kg st each)
Maximum speed	547mph at 31,170ft (880km/hr at 9,500m)
Range	1,400 miles (2,250km)
Weight	26,500lb (12,020kg) loaded
Span	68ft 10½in (21.0m)
Length	54ft 1in (16.48m)
Armament	Four cannon

Focke-Wulf Night Fighter II

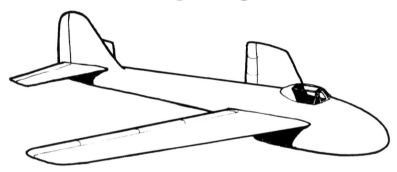

Details of the first in this series have been lost, but it probably set the design trend for the succeeding variants. Sketch designs for the *Nachtjäger* (Night Fighter) II, a night and bad-weather fighter, depict an aircraft of mid-wing layout with a conventional tail. The wings are sharply swept and the two HeS 011A turbojets are housed side by side in the lower nose section, while an elongated air intake gives the nose a shark-like appearance. The crew of three were to be accommodated in a pressurised cabin located above the engines and equipped with ejection seats. Armament consisted of four 30mm forward-firing cannon and two oblique, upward-firing 30mm cannon.

Focke-Wulf Night Fighter II data

Role	Three-seat jet night and bad-weather fighter
Ultimate status	Design
Powerplant	Two HeS 011A turbojets, 2,866lb (1,300kg) st each
Maximum speed	516mph (830km/hr) at sea level, 565mph at 22,970ft (910km/hr at 7,000m)
Service ceiling	42,650ft (13,000m)
Endurance	2hr 45min
Weight	27,280lb (12,375kg) loaded
Span	51ft 9½in (15.77m)

Focke-Wulf Night Fighter III

The third version in this series was simply a development of the previous design. Of lighter construction and with a smaller wing span and area, it was expected to demonstrate a higher speed and greater range. The twin-turbojet powerplant was located as in the Night Fighter II, but with the air intake placed centrally in the extreme nose.

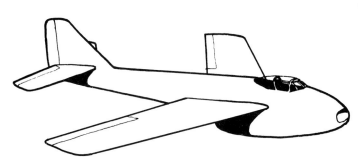

Focke-Wulf Night Fighter III data

Role	Three-seat jet night and bad-weather fighter
Ultimate status	Design
Powerplant	Two HeS 011A turbojets, 2,866lb (1,300kg) st each
Maximum speed	584mph at 19,690ft (940km/hr at 6,000m)
Endurance	2hr 48min
Weight	25,860lb (11,730kg) loaded
Span	46ft 3½in (14.10m)

FOCKE-WULF 011 SERIES

There is a long but poorly documented list of Focke-Wulf jet fighter projects, mostly powered by the HeS 011A turbojet and many bearing a strong resemblance to designs in the Ta 183 series. Details of this 011 series, as it was known, are so sparse that some of the following projects may in fact be described twice under different designations.

Focke-Wulf JP.011

Projected as a twin-tailboom fighter with a single HeS 011 turbojet, the JP.011 has been described as bearing a similarity to the de Havilland Vampire. It was also envisaged with a BMW 003 powerplant, and the sketch design based on this engine was mentioned in connection with the Ta 183 *Flitzer* project.

Focke-Wulf JP.011 data

Role	Single-seat jet fighter
Ultimate status	Design
Powerplant	One HeS 011A turbojet (2,866lb, 1,300kg st) or one BMW 003A turbojet (1,764lb, 800kg st)
Maximum speed	603mph at 29,530ft (970km/hr at 9,000m) (HeS 011), 478mph at 13,120ft (770km/hr at 4,000m) (BMW 003)
Range	652 miles (1,050km) (HeS 011)
Weight	8,003lb (3,630kg) (011), 6,945lb (3,150kg) (003) loaded
Span	26ft 3in (8.0m)

Focke-Wulf P.011-001

Another twin-tailboom project, the P.011-001 was basically similar to the JP.011 except for the addition of a Walter rocket booster.

Focke-Wulf P.011-001 data

Role	Single-seat mixed-power fighter
Ultimate status	Design
Powerplant	One HeS 011A turbojet (2,866lb, 1,300kg st) and one Walter 109-509 rocket motor (3,307lb, 1,500kg thrust)
Maximum speed	575mph at 28,870ft (925km/hr at 8,800m)
Range	373 miles (600km)
Span	26ft 3in (8.0m)

Focke-Wulf P.011-018a

Two versions of this single-seat fighter are known, one with a single 2,500hp Jumo 222 piston engine and the other with an HeS 011 turbojet.

Focke-Wulf P.011-018a data

Role	Single-seat jet fighter
Ultimate status	Design
Powerplant	One HeS 011A turbojet, 2,866lb (1,300kg) st
Maximum speed	597mph at 22,970ft (960km/hr at 7,000m)
Span	32ft 9½in (10.0m)

Focke-Wulf P.011-025

This single-seat fighter-bomber was to have two HeS 011 turbojets mounted in the fuselage and was proposed in three versions: fighter, long-range fighter and fighter-bomber. The wings were swept 40° and tailplane 42½°.

Focke-Wulf P.011-025 data

Role	Single-seat jet fighter, long-range fighter, fighter-bomber
Ultimate status	Design
Powerplant	Two HeS 011A turbojets, 2,866lb (1,300kg) st each
Maximum speed	572mph (920km/hr) at sea level, 668mph at 26,250ft (1,075km/hr at 8,000m), 647mph at 45,930ft (1,040km/hr at 14,000m)
Range	630 miles at 36,090ft (1,015km at 11,000m) (425 Imp gal, 1,930 lit fuel, fighter), 1,200 miles at 36,090ft (1,930km at 11,000m) (785 Imp gal, 3,570lit fuel, long-range fighter)
Weight	16,300lb (7,395kg) (fighter), 19,445lb (8,820kg) (long-range fighter), 19,800lb (8,980kg) (fighter-bomber) loaded
Span	41ft (12.50m)
Armament	Four 30mm cannon (+ 2,205lb, 1,000kg of bombs, fighter-bomber)

Focke-Wulf P.011-37a

Layout of this single-seat fighter project is unknown, but it was to be powered by a single 2,866lb (1,300kg) st HeS 011A turbojet. Wing span was 31ft 1½in (9.45m), anticipated maximum speed 603mph at 15,420ft (970km/hr at 4,700m), and operational range 609 miles (980km).

Focke-Wulf P.011-045

Again of unknown layout, the -045 was a large night fighter to be powered by two HeS 011A turbojets.

Focke-Wulf P.011-045 data

Role	Two-seat jet night fighter
Ultimate status	Design
Powerplant	Two HeS 011A turbojets, 2,866lb (1,300kg) st each
Maximum speed	560mph at 19,690ft (900km/hr at 6,000m)
Weight	27,008lb (12,250kg) loaded
Span	56ft 9in (17.30m)
Armament	Four 30mm cannon

Focke-Wulf P.021-009

Although it was not powered by a pure turbojet engine, the 021-009 is included because its powerplant was a turboprop adaptation of the HeS 011 turbojet. Once again, the layout is unknown, but the wing span was similar to that of the Ta 183 P.VIII HeS 021-powered fighter, indicating a possible relationship. A slower version, with heavier armament but otherwise similar, was also envisaged.

Focke-Wulf P.021-009 data

Role	Single-seat turboprop fighter
Ultimate status	Design
Powerplant	One HeS 021 turboprop, 2,400eshp
Maximum speed	560mph at 29,580ft (900km/hr at 9,000m)
Range	634 miles (1,020km)
Span	26ft 11in (8.20m)

Focke-Wulf BMW 003 fighter

This aircraft was derived from the Ta 183 project but was to be powered by the lower-rated BMW 003 turbojet of 1,764lb (800kg) st.

Focke-Wulf BMW 003 fighter data

Role	Single-seat jet fighter
Ultimate status	Design
Maximum speed	503mph at 19,690ft (810km/hr at 6,000m)
Weight	6,725lb (3,050kg) loaded
Span	24ft 7in (7.50m)
Armament	Two 30mm cannon

Gotha P.52 and P.53

These two designs were interim projects coming between the Horten Ho IX flying-wing fighter testbed and the Ho 229, later to become the Go 229. Power was to be supplied by two Jumo 004B turbojets of 1,984lb (900kg) st each.

Gotha P.60

There were three versions of the P.60 all-wing fighter design, each powered by a pair of turbojets mounted one above and one below the central wing section.

Powered by two BMW 003As, the P.60A was the cleanest of the three designs, with the pilot and observer lying prone behind nose glazing which formed the front of the fuselage. There was no fin and rudder, directional control being provided by aerofoils at each wingtip which could be retracted into the wing when not in use. The leading and trailing edges were swept back, the leading edge at 45°, and long flaps were fitted to improve stalling characteristics.

As the proposed successor to the Go 229, the P.60 was armed with four 30mm cannon and was to be boosted during take-off and climb by a 4,410lb (2,000kg) thrust Walter rocket motor.

The P.60B was larger than the first design and was expected to be significantly heavier. Performance was increased by the use of the more powerful HeS 011 turbojet. With jet engines and rocket operating, the P.60B was expected to attain 29,530ft (9,000m) from sea level in just 2min 36sec.

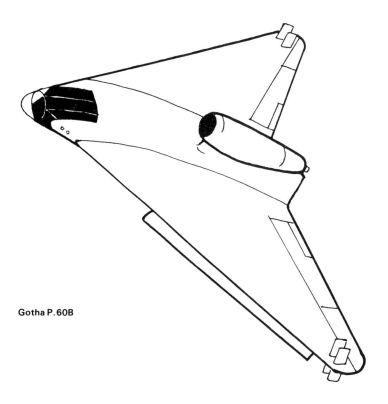

Gotha P.60B

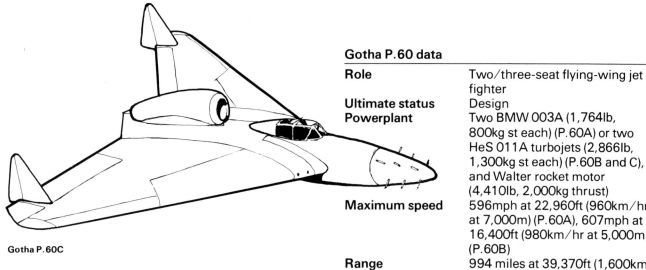

Gotha P.60C

The third version in the series, the P.60C, had the same basic form as the A and B but was intended as a bad-weather and night fighter. A more conventional nose and cockpit canopy were envisaged, with the crew of two or three sitting in tandem. A fin and rudder assembly was to be mounted on the outer trailing edge of each wing. The extended nose housed a *Morgenstern* radar antenna for the *Neptun* AI radar, and a pair of oblique upward-firing 30mm cannon supplemented the four forward-firing cannon.

Gotha P.60 data

Role	Two/three-seat flying-wing jet fighter
Ultimate status	Design
Powerplant	Two BMW 003A (1,764lb, 800kg st each) (P.60A) or two HeS 011A turbojets (2,866lb, 1,300kg st each) (P.60B and C), and Walter rocket motor (4,410lb, 2,000kg thrust)
Maximum speed	596mph at 22,960ft (960km/hr at 7,000m) (P.60A), 607mph at 16,400ft (980km/hr at 5,000m) (P.60B)
Range	994 miles at 39,370ft (1,600km at 12,000m) (P.60A), 1,647 miles at 39,370ft (2,650km at 12,000m) (P.60B and C)
Weight	16,390lb (7,435kg) (P.60A), 22,000lb (9,980kg) (P.60B) loaded
Span	40ft 8½in (12.40m) (P.60A), 44ft 4in (13.50m) (P.60B and C)
Length	31ft 6in (9.60m) (P.60A), 35ft 1½in (10.70m) (P.60B), 37ft 4½in (11.40m) (P.60C)

Gotha Go 229

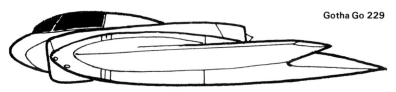

Gotha Go 229

Development of the Horten Ho IX as a twin-jet flying-wing fighter (see page 89) was transferred to the Gothaer Waggonfabrik in 1944 under the RLM designation 8-229. The Ho IX V1 was a glider, while the V2 was fitted with a pair of Jumo 004B-1 turbojets. Both flew in 1944, the V2 under power. The latter was destroyed very early in the programme, and a third prototype was built by Gothaer Waggonfabrik.

The first Gotha prototype (Go 229 V3) differed from the Horten versions in having a shallow bulge on the underside of the centre section and straight air intakes instead of the upswept arrangement used on the Ho IX. The aircraft took the form of a pure wing with tricycle undercarriage and a central cockpit section with a turbojet on either side; there were no vertical control surfaces.

The V3 was never flown, the factory having been occupied by Allied troops before it could be completed. The V4 and V5, intended as two-seat night fighter prototypes, were also near completion, and work on the V6 and V7 armament testbeds had started. Preparations for the production of an initial batch of 20 Go 229A-0 fighter-bombers were also under way.

Gotha Go 229 V3 minus outer wing panels (*via Pilot Press*)

Gotha Go 229A data

Role	Single-seat flying-wing jet fighter-bomber
Ultimate status	Flight test
Powerplant	Two Junkers Jumo 004C turbojets, 2,205lb (1,000kg) st each
Maximum speed	590mph (950km/hr) at sea level, 640mph at 21,320ft (1,030km/hr at 6,500m)
Range	1,180 miles (1,900km)
Ceiling	51,000ft (15,500m)
Weight	18,695lb (8,480kg) loaded
Span	54ft 11¾in (16.75m)
Length	24ft 6in (7.47m)
Armament	Four MK 103 or MK 108 30mm cannon and two 2,205lb (1,000kg) bombs

Gotha Go 345B

Designed originally as a troop-carrying glider, the Go 345A had a high wing and a single fin and rudder, and offered accommodation for ten fully equipped troops. Another glider version was to have a hinged nose section for use as a cargo transport.

The Go 345B was designed with two auxiliary Argus pulsejets, one under each wing. It was otherwise generally similar to the glider version but had a central skid instead of the tricycle wheeled undercarriage of the Go 345A.

The Go 345B did not go into production, though a prototype was completed in 1944.

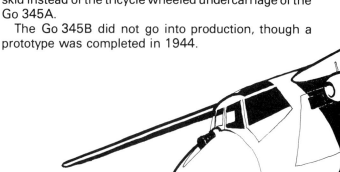

Gotha Go 345B data

Role	Pulsejet-powered troop-carrying glider
Ultimate status	Construction
Powerplant	Two Argus As 014 pulsejets, 770lb (350kg) thrust each
Maximum speed	192mph (310km/hr)
Weight	13,230lb (6,000kg) loaded
Span	68ft 9in (20.96m)

Heinkel Strabo 16

Designed in 1944 as a mid-wing fast bomber, the Strabo (from *Strahlbomber*, jet bomber) 16 was to be powered by four Junkers Jumo 004C turbojets. Its general layout and fate are unknown.

Heinkel Strabo 16 data

Role	Fast jet bomber
Ultimate status	Design
Powerplant	Four Junkers Jumo 004C turbojets, 2,205lb (1,000kg) st each
Maximum speed	535mph (855km/hr)
Take-off weight	35,280lb (16,000kg)
Span	59 ft 1in (18.0m)
Length	54ft 3½in (16.54m)

Heinkel P.135

There is even less information about the P.135, which is referred to as a fighter with swept wings and possibly powered by turbojet engine(s).

Heinkel He 162A Salamander (Volksjäger)

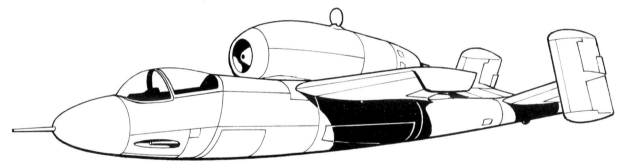

On September 8, 1944, the RLM issued a specification for a high-speed *Volksjäger* (People's Fighter) which called for an aircraft with a weight of no more than 4,410lb (2,000kg) and over 30min endurance. Apart from a BMW 109-003 turbojet, a minimum of strategic materials was to be used, and the design was to be ready for mass production by semi-skilled labour on January 1, 1945. Armament was to consist of two 30mm cannon and the aircraft was to be able to take off within a distance of 1,640ft (500m).

Five manufacturers submitted designs. Heinkel in fact submitted two, but was clearly pinning its hopes on the P.1073, as evidenced by the full-scale mock-up produced for inspection by the RLM. This probably carried the day for Heinkel, and the P.1073 was ordered into production as the He 500 (later changed to He 162).

The most remarkable aspect of the He 162's history is the brief interval between the issue of the official specification on September 8, 1944, and the flight of the first prototype on December 6, 1944. By February 1945 production aircraft were being delivered to *Luftwaffe* units. In all, some 116 examples of the *Salamander* (earlier named *Spatz*, Sparrow) were built before the war ended. This was a truly remarkable achievement, and while some of the rival designs in the *Volksjäger* competition were technically superior, their sophistication would probably have resulted in an unacceptably protracted development programme.

Layout was very simple, with straight shoulder wings and short anhedral wing tips. The engine was located on top of the fuselage, just behind the cockpit. The tailplane had sharp dihedral, with twin, inward-canted fins at the extremities.

Several variations of the He 162 were envisaged,

Above: He 162A in Luftwaffe markings (*Archiv B. Holsen*)

Below: He 162A in British markings after capture (*IWM*)

mostly centring on changes of powerplant. One of the earliest of these was a proposal to replace the basic BMW 003 turbojet with a BMW 003R which incorporated a 2,750lb (1,250kg) thrust rocket motor for use as a climb booster. This combination increased maximum speed to 628mph (1,010km/hr) at sea level and rate of climb to as much as 20,700ft/min (6,300m/min). A specially rated BMW 003E capable of giving increased thrust for up to 30sec was also envisaged for the He 162. The increase brought thrust up to 2,020lb (915kg) at sea level, enabling the He 162 to achieve a maximum speed of 552mph (890km/hr). Use of a single Junkers Jumo 004D or E, developing 2,205lb (1,000kg) at sea level, was also proposed, giving the aircraft a maximum speed similar to that of the BMW 003E-powered version, but for sustained periods.

Heinkel He 162A data

Role	Single-seat jet fighter
Ultimate status	Pre-operational
Powerplant	One BMW 003A turbojet, 1,764lb (800kg) st
Maximum speed	522mph at 19,690ft (840km/hr at 6,000m)
Range	410 miles at 36,100ft (660km at 11,000m)
Weight	3,483lb (1,580kg) empty, 5,940lb (2,695kg) loaded
Span	23ft 7¾in (7.20m)
Length	29ft 8½in (9.05m)
Wing area	120.56ft^2 (11.2m^2)
Armament	Two MG 151/20 20mm cannon with 120 rounds per gun

Heinkel He 162B-1 and B-2

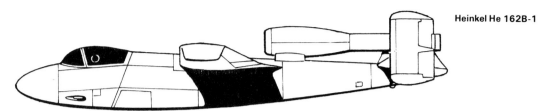

Heinkel He 162B-1

Two months after the RLM issued its *Volksjäger* specification in September 1944, a new specification calling for a *Miniaturjäger* (Miniature Fighter) went out to the German aircraft industry. The requirement was for a simple high-speed fighter which could be built even more rapidly than the *Volksjäger* but which was not to be semi-expendable like the Bachem *Natter*. Three manufacturers – Blohm und Voss, Heinkel and Junkers – presented designs, and all three had come to the conclusion that the time needed to build jet engines was the biggest obstacle to high-speed production. Since a pulsejet engine took some 450 man-hours less than a turbojet to build, all three designs were to be powered by very simple pulsejets of the type fitted to the V-1 flying bomb.

The Heinkel design was no more than a modified He 162 *Salamander* and was to be known as the He 162B. Two versions were envisaged: the B-1 with two Argus 109-014 pulsejets, and the B-2 with a single, more powerful Argus 109-044 unit. Like the *Volksjäger*, they were to have the engines mounted dorsally, but in the case of the *Miniaturjäger* the pulsejets were set much further back. This was probably done to reduce some of the very damaging acoustic effects produced by these engines and first encountered on the earlier Me 328 prototypes.

The result of the competition is not known, though the Junkers design (EF.126) reached the mock-up stage.

Heinkel He 162B data

Role	Single-seat pulsejet fighter
Ultimate status	Design
Powerplant	Two Argus 109-014 pulsejets (734lb, 330kg thrust each) (B-1), one Argus 109-044 pulsejet (1,112lb, 500kg thrust) (B-2)
Maximum speed	503mph (810km/h) at sea level (B-1), 441mph (710km/hr) at sea level (B-2) 441mph at 20,000ft (710km/hr at 6,100m) (B-1), 376mph at 20,000ft (605km/hr at 6,100m) (B-2)
Range	255 miles at 20,000ft (410km at 6,100km) (B-1), 236 miles at 20,000ft (380km at 6,100m) (B-2)
Service ceiling	26,000ft (7,900m) (B-1), 21,320ft (6,500m) (B-2)
Climb rate	3,650ft/min (1,110m/min) (B-1), 2,360ft/min (720m/min) (B-2)
Weight	7,260lb (3,290kg) loaded (B-1), 6,380lb (2,895kg) loaded (B-2)
Span	23ft 7¾in (7.20m)
Length	29ft 8½in (9.05m)

Heinkel He 162C and D

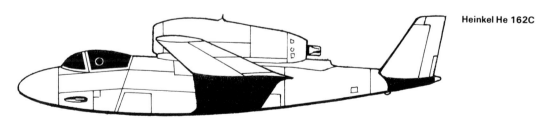

Heinkel He 162C

The final development of the He 162 design was a swept-wing version powered by a single HeS 011 turbojet exhausting between the surfaces of a new butterfly tailplane. The effects of wing sweep were still very much under investigation at this time and two variations were proposed: the C with forward-swept

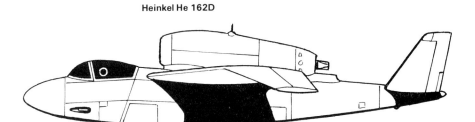

Heinkel He 162D

wings and the D with 38° rear-swept wings. An initial prototype, half completed by the end of the war, was to be capable of taking both types of wing.

This combination of the more powerful HeS 011 turbojet and swept wings was expected to boost the He 162's maximum speed to over 570mph (920km/hr). Production was intended to begin soon after the prototype had been tested in 1946.

Heinkel He 162C/D data

Role	Single-seat jet fighter
Ultimate status	Construction
Powerplant	One HeS 011A turbojet, 2,866lb (1,300kg) st
Maximum speed	572mph at 22,960ft (920km/hr at 7,000m)
Weight	8,050lb (3,650kg) loaded
Span	26ft 3in (8.0m)
Length	30ft 4in (9.25m)
Armament	Two MK 103 30mm cannon in limited-elevation mounts

Heinkel He 176

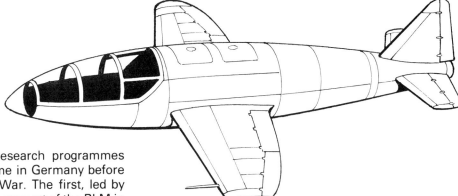

Two full-scale rocket aircraft research programmes were under way at the same time in Germany before the start of the Second World War. The first, led by Alexander Lippisch, received the support of the RLM in 1937. Code-named *Projekt* X, it eventually led to the production of the Me 163 *Komet* (see DFS, page 43).

The other programme was undertaken privately by Ernst Heinkel and was as energetically pursued as the official venture. In fact, Heinkel was investigating the possibilities of both rocket and turbojet engines at the same time, with the rocket-powered He 176 being built alongside the turbojet-powered He 178 in the same hangar.

Early rocket experiments had been performed with various aircraft and engines, and after successful tests with a He 112 and Walter liquid-fuel rockets, Heinkel decided to build an exclusively rocket-powered aircraft for high-speed research. Aiming for a maximum speed of 620mph (1,000km/hr), design work on the He 176 started in 1937. The very small fuselage was to house the pilot, undercarriage, two tanks for the hydrogen peroxide and methanol propellants, and a Walter R.I (TP-2) variable-thrust rocket motor (100-1,102lb, 45-500kg thrust) with an endurance of 60sec at maximum thrust.

Two designs seem to have existed, the cleaner of which had an enclosed cockpit with glazing over the

He 176, open-cockpit version (*via Pilot Press*)

upper front fuselage and nose section. This has been illustrated in drawing form only. The other version was similar but cruder, with a conventional open cockpit.

The first flight was made on June 20, 1939, followed the next day by another 50sec flight. Though these sorties were incident-free it was obvious that the wing was barely capable of supporting the aircraft in flight. On July 3 the He 176 was flown before Hitler, Goering and von Keitel, but the RLM had already decided that there was more promise in its own *Projekt* X, and the Heinkel 176 was dropped at the start of the war.

Heinkel He 176 data

Role	High-speed rocket research aircraft
Ultimate status	Flight test
Powerplant	One Walter HWK R.I-203 rocket motor, 1,323-1,521lb (600-690kg) thrust
Maximum speed	466mph at 13,120ft (750km/hr at 4,000m)
Climb	13,120ft (4,000m) in 1min 6sec
Ceiling	29,530ft (9,000m)
Weight	1,720lb (780kg) empty equipped, 3,572lb (1,620kg) maximum loaded
Span	16ft 5in (5.00m)
Length	17ft 1in (5.20m)
Wing area	58.13ft² (5.4m²)

Heinkel He 178

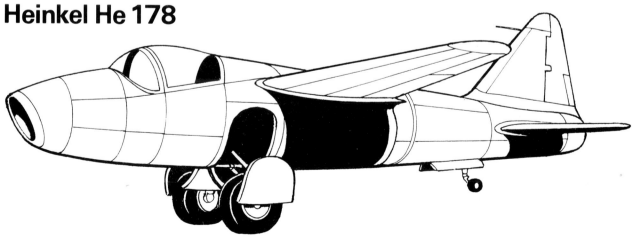

Early in 1936 Ernst Heinkel engaged Dr Pabst von Ohain and Max Hahn to continue their own line of jet propulsion research and development. Their first engine, the hydrogen-powered S 1 radial turbine, was completed in the same year as a demonstration model. This was followed by the HeS 2, which also used hydrogen. First bench-tested in March 1937, the HeS 2 achieved 287lb (130kg) thrust at 10,000rpm.

Soon afterwards it was decided to change over to petrol fuel and during the following months the novel powerplant was completely redesigned, emerging as the HeS 3. First bench-tested early in 1938, it developed 992lb (450kg) thrust and proved much more controllable.

Later in 1938 flight tests were carried out with another engine, the axial-flow HeS 3A, mounted beneath an He 118. This engine was test-flown many times until the turbine finally burned out. By this time the programme had advanced sufficiently for an aircraft to be designed around a turbojet engine. The result was a marriage of the improved HeS 3B and the He 178 airframe, which was built alongside the rocket-powered He 176 during 1938.

The He 178 had a straight shoulder-wing layout, with the single turbojet mounted in the rear fuselage and fed by an air intake in the extreme nose. The undercarriage was of the conventional tailwheel type.

By August 1939 the He 178 was making short hops, and a first true flight was made on August 27, though this was cut short after the aircraft suffered the world's first jet birdstrike. The He 178 was thus the very first exclusively turbojet-powered aircraft to fly, preceding Britain's Gloster E.28/39 by 21 months.

Throughout its development the He 178 remained a private venture, and RLM officials did not get to see it until November 1939. But the war had started by this time and they decided to concentrate their efforts on conventional aircraft production. Nevertheless, Heinkel persisted with its development, which led to the 1,300lb (590kg) st HeS 6 engine. This unit was flight-tested but its massive weight of 925lb (420kg) resulted in disappointing performance.

Flight tests of the He 178 were plagued by problems with the undercarriage, which repeatedly failed to retract. Possibly as a result of this, the He 178 never flew faster than 373mph (600km/hr), even though von Ohain believed that a speed of 530mph (850km/hr) was possible with the HeS 6 installed. Development was finally cut short to make way for the He 280 twin-jet fighter.

Heinkel He 178 data

Role	Single-seat jet research aircraft
Ultimate status	Flight test
Powerplant	One HeS 3B turbojet (1,102lb, 500kg st); one HeS 6 (1,300lb, 590kg st) proposed
Maximum speed	373mph (600km/hr) (up to 530mph, 850km/hr estimated)
Weight	3,565lb (1,616kg) empty, 4,396lb (1,995kg) loaded
Span	23ft 7¾in (7.20m)
Length	24ft 6½in (7.48m)
Wing area	97.95ft² (9.1m²)

Heinkel He 280

Having flown the world's first jet-powered aircraft fully 20 months before its nearest rival, the Gloster E.28/39, Heinkel rubbed salt into the wound by also flying the first turbojet-powered fighter nearly six weeks before the British aircraft took to the air. The warplane was the He 280, on which development work had started back in 1939. This single-seater was not only unique at that time in having two jet engines, but it was also fitted with the world's first ejection seat, operated by compressed air, and had a tricycle undercarriage, a rare feature at that time.

The wings were straight and mid-set, with the turbojets slung beneath at about one-third span. A long, pointed nose led back to the single-seat pressurised cabin, and the high-set tailplane carried twin endplate fins.

Starting in September 1940, over 40 glide tests were carried out with the first prototype, and in March 1941 two Heinkel-Hirth HeS 8 turbojets of 1,322lb (600kg) st each were fitted. The first powered flight, with the engine cowlings removed to minimise a fire hazard that had shown up in ground tests, took place on March 30, 1941. Eight He 280 prototypes were eventually built, and one of them was pitted against an FW 190 piston-engined fighter in a mock battle late in 1941 to prove the superiority of jet-powered aircraft. But even this successful display failed to persuade the RLM to sanction mass production, and the project was finally wound up.

The first of the He 280 prototypes did however enjoy a short revival early in 1942 when it underwent a radical change of powerplant. Six 770lb (350kg) thrust Argus As 014 pulsejets were fitted beneath the wings and the aircraft was used to test-fly the engine which was eventually to power the Fieseler Fi 103 flying bomb. The aircraft was towed into the air on January 13, 1942, by a pair of Bf 110s. But when it was released at a little under 8,000ft (2,400m) it was found to have iced up and the pilot was forced to make the first emergency ejection from an aircraft.

Heinkel He 280 V7 (via Alex Vanags)

Heinkel He 280 data*

Role	Single-seat jet fighter
Ultimate status	Flight test
Power plant	Two HeS 8A turbojets (1,587lb, 720kg st each) (four prototypes), or two BMW 003A (1,675lb, 760kg st each), or two Jumo 004As (1,850lb, 840kg st each), or two Jumo 004Bs (1,984lb, 900kg st each)
Maximum speed	467mph (750km/hr) at sea level, 508mph (818km/hr) at 19,690ft (6,000m)
Range	382 miles (615km)
Service ceiling	37,400ft (11,400m)
Weight	11,475lb (5,205kg) loaded
Span	39ft 4½in (12.0m)
Length	33ft 5½in (10.20m)
Wing area	231.5ft^2 (21.5m^2)
Armament	Three MG 151/20 20mm cannon

*With Jumo 004B-1.

He 280 about to land after jet-powered test flight (*via Pilot Press*)

Heinkel P.1063

Poorly documented, the single-seat P.1063 was projected in 1942 as a mid-wing bomber powered by two turbojets.

Heinkel P.1068

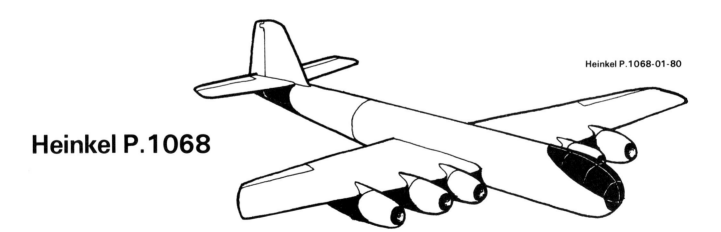

Heinkel P.1068-01-80

P.1068 was the original prototype designation of the He 343 four-jet bomber. Projected in 1944, a series of four variations on the same basic theme were proposed. All were of a mid-wing layout, but with varying numbers and locations of powerplant.

The first in the series, the P.1068-01-78, had straight wings and was to be powered by four 2,866lb (1,300kg) st HeS 011A turbojets. Wing span and length were 62ft 4in (19.0m) and 65ft 7in (20.0m) respectively, take-off weight 49,160lb (22,300kg) and maximum speed 530mph (850km/hr).

The next in the series, the P.1068-01-80, was basically similar to the -78 but had two more HeS 011s, the resulting total of six engines being positioned in threes spaced evenly beneath each wing. Take-off weight was consequently increased to 51,816lb

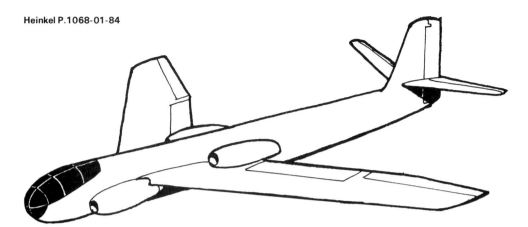

Heinkel P.1068-01-84

(23,504kg) and maximum speed to 578mph (930km/hr). The fuselage was narrow, with a glazed nose enclosing the crew cabin. The wings were straight and the tailplane surfaces were set on either side of the base of the single fin.

Third in the series was the P.1068-01-83, which again had a straight mid-wing layout and was powered by four HeS 011s. This design was slightly smaller than the previous two, with a wing span of 52ft 6in (16.0m) and a length of 55ft 10in (17.0m). Take-off weight, at 39,594lb (17,960kg), was also down. This all-round loss of bulk was expected to result in an increase in maximum speed to 565mph (910km/hr).

The final design in the series, the P.1068-01-84, differed radically from the previous three. While the fuselage was basically similar, the wings were swept at a pronounced 45° and the four HeS 011s were located in twos, the forward pair on either side of the nose, just below and in front of the wing leading edges, and the second pair further back and above the wing trailing edges. Wing span remained 52ft 6in (16.0m) and length was 58ft 9in (17.90m). Take-off weight was increased to 40,250lb (18,255kg) but, surprisingly, maximum speed was put at only 556mph (895km/hr), less than that of the previous, straight-wing version.

It is not clear which of these four designs was chosen for development, but the DFS acquired and adapted the project in 1944 for its own purposes, using it to study various wing-sweep angles. Five versions were to be built, all powered by rocket engines and fitted with dummy turbojet nacelles, but the project suffered a terminal setback when the nearly complete first aircraft was destroyed by fire.

Heinkel He 343

Heinkel He 343A-2

A direct progression from the P.1068 projects, the He 343 was intended as the production version and was proposed in at least four variants designated He 343A to D, plus sub-versions.

The basic He 343 was a straight mid-wing aircraft with a pair of separated turbojets beneath each wing. The fuselage and tail were similar to those proposed for the P.1068, and a glazed nose accommodated the two-man crew. In all, it bore a superficial resemblance to the Arado Ar 234, from which it is said to have been developed.

The He 343A, designed in 1944, was the first and main version. There were at least four variations: the A-1 heavy fighter or fast bomber, the A-2 reconnaissance aircraft or fast bomber, and the A-3 and A-3/J heavy fighters. All were to be powered by either the Jumo 004C or HeS 011A turbojets.

The He 343B was to be similar to the A-1, and the He 343C was a variation of the A-2 with extra fuel capacity. The D was a further variation of the reconnaissance C and was powered by four 2,866 (1,300kg) st HeS 011s.

Heinkel 343 data

Role	Two-seat bomber, reconnaissance aircraft or heavy fighter (data for the He 343D reconnaissance variant)
Ultimate status	Design
Powerplant	Four HeS 011A turbojets, 2,866lb (1,300kg) st each
Maximum speed	572mph (920km/hr) at sea level
Range	603 miles (970km)
Service ceiling	41,670ft (12,700m)
Weight	40,000lb (18,145kg) loaded
Span	58 ft 9½in (17.92m)
Length	55ft 9in (17.0m)
Wing area	455ft² (42.27m²)
Armament	Two MG 151/20 20mm cannon (4,410lb, 2,000kg of bombs on bomber versions)

Heinkel P.1069

Projected in 1943 as a mid-wing jet fighter, the P.1069 was to be powered by a single Jumo 004B turbojet of 1,984lb (900kg) st.

Heinkel P.1070

Another 1943 design study was the P.1070 flying wing, which was to be powered by two or four 1,984lb (900kg) st Junkers Jumo 004B turbojets and was possibly intended as a research aircraft.

Heinkel P.1071

This unusual fighter project, also designed in 1943, had an asymmetric layout and was to be powered by two piston engines or a pair of Jumo 004B turbojets.

Heinkel P.1072

Again designed in 1943, this project was to be a long-range fast bomber. It had a straight mid-wing layout and was to be powered by either four piston engines or four 1,764lb (800kg) st BMW 003 turbojets.

Heinkel P.1073

Heinkel seemed to indulge more than most in the confusing habit of using prototype designations for more than one project, or at least using one designation for several different designs within the same project. One such is P.1073, originating in February 1944. It was later given as the original designation of the He 162 and of later versions of the He 162 using a Jumo 004C or HeS 011 turbojet (P.1073/II). But the first aircraft to which this designation was applied was the P.1073.01 (P.1073/I). Power was to be supplied by

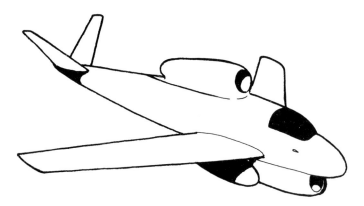

two HeS 011 or two Jumo 004C turbojets mounted one above the fuselage in a similar fashion to the He 162 installation, and one below and slightly offset to the right of the nose. The fuselage was similar to that of the He 162, but the mid-set wings were swept and carried a large, streamlined 110 Imp gal (500lit) fuel tank beneath each inner leading edge, and the tail surfaces were swept up in a butterfly arrangement.

The development history of the P.1073/I is not clear, but it seems to have been dropped at an early stage. The designation was then revived for Heinkel's *Volksjäger* proposal in September 1944, possibly in an effort to confuse the commercial opposition or the enemy.

Heinkel P.1073 data

Role	Single-seat jet fighter (data for P.1073.01-04 with HeS 011s)
Ultimate status	Design
Powerplant	Two HeS 011A turbojets, 2,866lb (1,300kg) st each
Maximum speed	627mph (1,010km/hr)
Weight	13,500lb (6,125kg) loaded
Span	39ft 4in (12.0m)
Length	33ft 8in (10.26m)
Armament	Two MG 151/20 20mm cannon

Heinkel P.1077 (Julia)

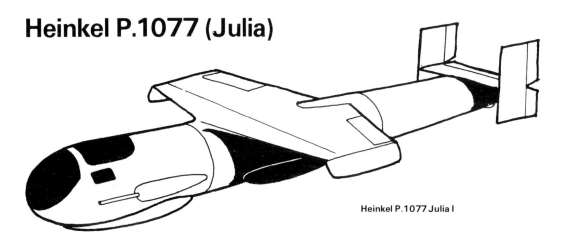

Heinkel P.1077 Julia I

Designed to meet the same basic requirements as the Bachem *Natter* target-defence interceptor, the P.1077 prototype did not reach the flight-test stage, although a mock-up and two airframes were approaching completion when discovered by Allied forces in 1945.

Design work had started in July 1944 and led to the projecting of three basic versions. The first, named Julia I, accommodated the pilot in a prone position in the nose. Two 30mm cannon were mounted on either side of the cabin and aimed by a simplified AA gunnery-control radar. The main powerplant was to be a single 3,750lb (1,700kg) thrust Walter rocket motor, assisted during the near-vertical take-off by four 2,646lb (1,200kg) thrust jettisonable booster rockets. The Julia I was expected to reach a target at 39,370ft (12,000m) in just one minute on full power, after which it could fly for another five minutes on the main Walter rocket motor's cruising chamber. The output of the cruising chamber could be varied between 330 and 661lb (150-300kg) thrust, giving a maximum range of 40 miles (65km) at 497mph (800km/hr) and 32,800ft (10,000m). After a glide approach the aircraft was to land on retractable skids which were faired into the fuselage during flight.

A second version, the Julia II, was basically similar

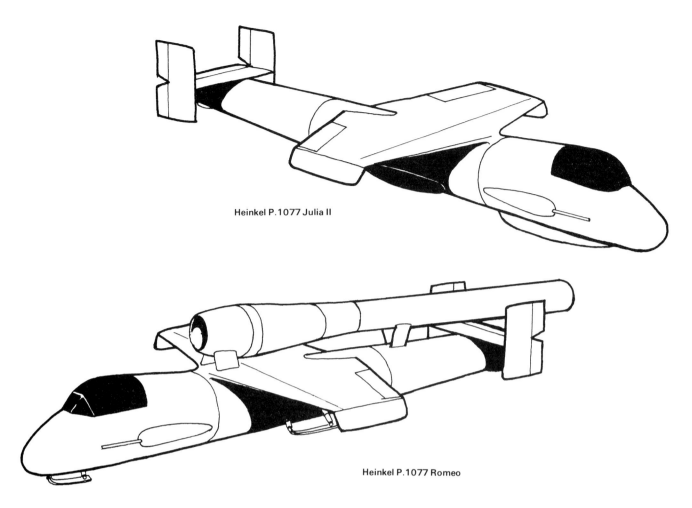

Heinkel P.1077 Julia II

Heinkel P.1077 Romeo

apart from the location of the pilot, who sat on a conventional seat under a slightly more arched canopy.

Basic layout for all three versions was a straight shoulder wing with anhedral tips, and an equally high-set tailplane with twin endplate fins.

Last of the P.1077 designs had a different powerplant and was re-named Romeo. The rocket motor was replaced by an Argus 014 pulsejet mounted on the back of a Julia II-type fuselage. Although many references give the Romeo as being designed, like the Julia, for the interceptor role, it was obviously intended for fast, cheap production in response to the *Miniaturjäger* requirement. The landing skids were located in the same way as those of the Julia, but they were of cruder construction and lacked the flush fairings. Performance of the Romeo was supposed to be comparable with that of its more powerful sisters, but this seems rather unlikely, especially as pulsejet efficiency drops off rapidly with height. A maximum speed of around 450mph (725km/hr) at sea level therefore seems more probable.

Heinkel P.1077 data

Role	Target-defence interceptor
Ultimate status	Construction
Powerplant	One Walter HWK 109-509A-2 rocket motor (3,750lb, 1,700kg thrust) and four jettisonable, 10sec-endurance solid-fuel booster rockets (2,646lb, 1,200kg thrust each) (Julia I and II)
	One Argus As 014 pulsejet (1,050lb, 480kg thrust) (Romeo)
Maximum speed	610mph (980km/hr) (Julia I)
Weight	3,950lb (1,790kg) loaded, with booster rockets (Julia I)
Span	15ft 1in (4.60m)
Length	22ft 3½in (6.80m)
Wing area	77.5ft² (7.2m²)
Armament	Two MK 108 30mm cannon

Heinkel P.1078

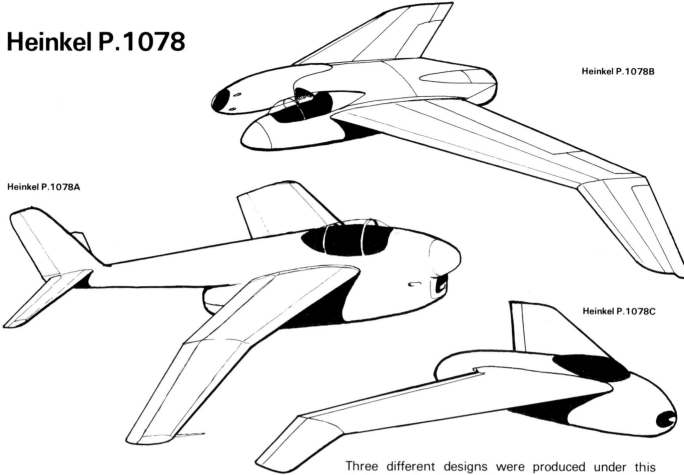

Heinkel P.1078B
Heinkel P.1078A
Heinkel P.1078C

Heinkel P.1078 data

Role	Single-seat jet fighter
Ultimate status	Design
Powerplant	One HeS 011A turbojet, 2,866lb (1,300kg) st
Maximum speed	609mph (980km/hr (A), 636mph (1,025km/hr) (B), 624mph (1,005km/hr) (C), all at sea level
Range	932 miles (1,500km) at 36,090ft (11,000m) (A), 960 miles (1,545km) at 36,090ft (11,000m) (B)
Service ceiling	42,300ft (12,900m) (A), 45,000ft (13,700m) (B)
Weight	8,906lb (4,040kg) (A), 8,576lb (3,890kg) (B), 8,643lb (3,920kg) (C)
Span	28ft 11in (8.80m) (A), 30ft 9½in (9.38m) (B), 29ft 6¾in (9.0m) (C)
Length	31ft 1½in (9.48m) (A), 20ft (6.10m) (B), 20ft (6.10m) (C)
Wing area	182ft² (16.9m²) (A), 219ft² (20.3m²) (B), 191.6ft² (17.8m²) (C)
Armament	Two MK 108 30mm cannon

Three different designs were produced under this designation for the OKL Emergency Fighter competition of 1944, the last of them being submitted for approval (see FW Ta 183, page 57).

The P.1078A had 40°-swept shoulder-mounted gull wings and a fuselage similar to that of the Messerschmitt P.1101, which had a single turbojet located in the lower centre section and exhausting below a central tailboom. The single-seat cockpit was placed well forward, with single 30mm cannon on either side.

Altogether different was the B, which was to be tailless and with no vertical control surfaces. The wings were swept at 40° and gulled, lateral control being provided by the anhedral wingtips. The fuselage had twin nose cones, the air intake for the single turbojet being set between and to the rear of them. The armament of two 30mm cannon was housed in the starboard nose cone, with the pilot seated in the other.

It was the final version, the P.1078C, which was submitted for the OKL competition. Like the B, it was a flying wing without vertical control surfaces. The shoulder-set wings were swept back at 40°, had anhedral wingtips, and contained the whole fuel supply in the inner sections. A single 2,866lb (1,300kg) st HeS 011 turbojet was located in the short, centreline fuselage, with the air intake in the extreme nose. The air intake was flattened so that the cockpit, with a 30mm cannon on either side, could be superimposed.

Although well thought out, the P.1078C was passed over in favour of the Focke-Wulf Ta 183.

Heinkel P.1079

This twin-jet night fighter series was designed in five forms, all under the same designation. The first was of mid-wing layout, with 35° of sweepback and V-tail. The crew of two sat back-to-back in a cockpit near the nose, and the two HeS 011A (or Jumo 004B) turbojets were mounted next to the fuselage in the wing roots.

The P.1079B bore a strong resemblance to the first in the series, though the tail and rear fuselage had been deleted to result in a flying-wing layout. A single vertical fin replaced the V-tail of the P.1079A, and the wings were gulled and at 45° more sharply swept. The seating for the two-man crew was staggered, permitting the nose to be shortened. A second version of the P.1079B approximated more closely to a true flying wing, the fin being omitted. The gulling of the wing was more pronounced and the wingtip anhedral increased. The engines were more widely spaced than in the previous designs, allowing the main undercarriage to be accommodated between powerplant and fuselage.

For the last three variants in the series the Heinkel designers reverted to a more conventional tailed layout. All were intended as night fighters except for the P.1079C, which was designed for reconnaissance duties and had 45°-swept wings. The night-fighter version (A-1) was to carry centimetric AI radar. The D and E had wings with 35° sweepback but otherwise their layout is unknown.

Heinkel P.1079 data

Role	Two-seat jet night fighter
Ultimate status	Design
Powerplant	Two HeS 011A turbojets, 2,866lb (1,300kg) st or (two Jumo 004B)
Maximum speed	584mph (940km/hr) (A), 630mph (1,015km/hr) (B), 615mph (990km/hr) (C), 590mph (950km/hr) (D), 584mph (940km/hr) (E), all at 22,960ft (7,000m)
Range	1,678 miles at (2,700km) at 36,090ft (11,000m) (A)
Span	41ft 8in (12.70m) (A and B), 36ft 1in (11.0m) (C), 39ft 4in (12.0m) (D), 48ft 6in (14.78m) (E)
Length	46ft (14.02m) (A), 29ft 6in (9.0m) (B), 48ft 5in (14.75m) (C), 43ft 8in (13.30m) (D), 50ft 2½in (15.30m) (E)
Armament	Four MK 108 30mm cannon (three MK 108 30mm cannon in Jumo-powered variants)

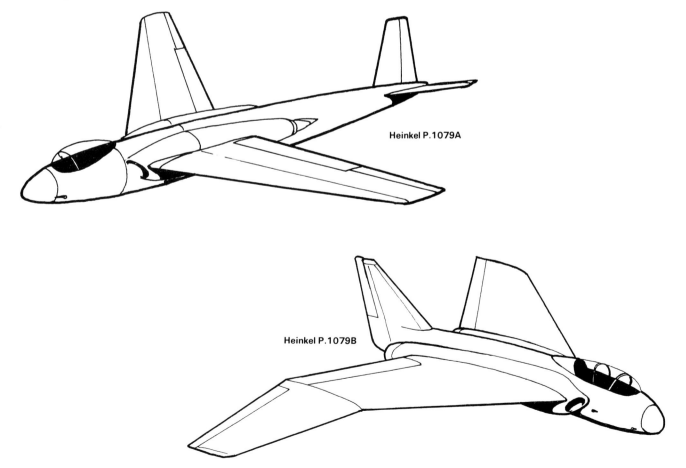

Heinkel P.1079A

Heinkel P.1079B

Heinkel P.1080

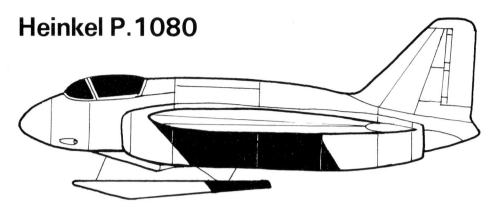

Experiments with ramjets started as early as 1937 in Germany, but until 1943 very little interest had been shown officially. Thereafter the effort enjoyed government support, and in November 1944 it was decided to apply DFS ramjet data to the development of a ramjet-powered Heinkel aircraft. The war was nearly over when Heinkel received the information and with it a request that provision be made for use of the "foam coal" fuel with which Professor Lippisch had been experimenting.

A flying-wing layout was chosen, using wings based on those intended for the P.1078C to save development time. There was a single fin and the single-seat cockpit was located well forward in the nose. Two massive Sänger-type ramjets, 16ft 5in long and nearly 3ft 4in in diameter at the combustion chamber, were placed at the wing roots on either side of the fuselage. Four jettisonable 2,205lb (1,000kg) rockets were to be used to take off and attain ramjet operating speed. Fuel for the very thirsty ramjets was located in the central fuselage, making it virtually a flying fuel tank from the cockpit back.

The project was initiated at too late a stage for development to have progressed very far before being halted by the end of the war.

Heinkel P.1080 data

Role	Single-seat ramjet fighter
Ultimate status	Design
Powerplant	Two Sänger-type ramjets (926lb, 420kg thrust each at sea level, up to 3,440lb, 1,560kg thrust each) and four take-off rockets (2,205lb, 1,000kg thrust each)
Maximum speed	311mph (500km/hr) at sea level, up to 621mph (1,000km/hr) maximum
Weight	9,500lb (4,300kg) loaded
Span	29ft 3½in (8.92m)
Length	26ft 9in (8.15m)
Wing area	215.2ft² (20.0m²)
Armament	Two MK 108 30mm cannon

Heinkel P.1090

This swept-wing fighter project is poorly documented. It is thought to have been proposed with two turbojet engines.

Heinkel Wespe

Referred to simply as the Heinkel *Wespe* (Wasp), this project was designed in 1945. Described as a "circular-wing aircraft", the *Wespe* was to be powered by a single 2,400eshp/1,740lb st Heinkel HeS 021 turboprop and was intended as a vertical take-off fighter. With this in mind, the term "circular wing" could be interpreted as meaning rotary wing, as in the Focke-Wulf *Triebflügel* (see page 60).

Heinkel Wespe data

Role	Vertical take-off turboprop fighter
Ultimate status	Design
Powerplant	One 2,400eshp/1,740lb (790kg) st HeS 021 turboprop
Maximum speed	497mph (800km/hr)
Weight	4,696lb (2,130kg) loaded
Span	16ft 4½in (5.0m) (probably means the diameter of the rotary wing)
Length	20ft 4in (6.20m)

Henschel P.122

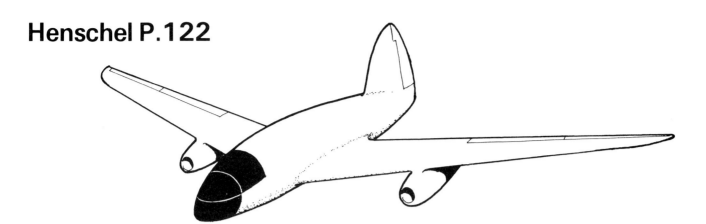

Not to be confused with the Hs 122, which was a single piston-engined reconnaissance aircraft, the P.122 was to be a high-altitude jet bomber. It was a tailless design with a single fin and moderately swept-back wings. The two powerful BMW 018 turbojets were slung beneath the low-set wings and gave the aircraft an estimated maximum speed of over 620mph (1,000km/hr) at sea level.

Henschel P.122 data

Role	High-altitude two-seat jet bomber
Ultimate status	Design
Powerplant	Two BMW 109-018 turbojets, 7,496lb (3,400kg) st each
Maximum speed	627mph (1,010km/hr) at sea level, 584mph at 32,800ft (940km/hr at 10,000m)
Range	1,243 miles at 55,770ft (2,000m at 17,000m)
Weight	33,290lb (15,000kg) loaded
Span	73ft 9in (22.48m)
Length	40ft 8 in (12.40m)

Henschel Hs 132

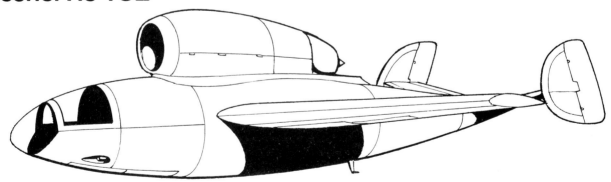

This aircraft was designed exclusively for ground attack duties to replace the vulnerable Ju 87 *Stuka*. As such, the new type had to have enough speed to be able to survive against enemy ground and air defences. The pilot was to lie prone behind a nose canopy, this posture being expected to help counter the effects of the high g forces encountered in the pull-out from the dive. Steep dives were not envisaged, so dive brakes were not to be fitted. As in the He 162, the turbojet was to be mounted on top of the fuselage to facilitate speedy construction and maintenance.

The all-wooden wings were straight and mid-set on the fuselage, which like the remainder of the aircraft was made of metal. The structure was designed to

Henschel Hs 132 V1 (*via Pilot Press*)

withstand a peak force of some 12g during the escape from an attack.

Six prototypes were ordered and construction began in March 1945. By the end of the war Hs 132 V1 was nearly complete, with the others in an advanced state. They were captured by the Russians and it is not known whether they were ever flown, although they were scheduled by Henschel to fly in June 1945.

Three versions were envisaged, with varying power-plants and armament. The Hs 132A was to have a BMW 003 turbojet and be able to carry one 1,100lb bomb, while the B was to have a slightly more powerful Jumo 004 and be armed with a 1,100lb bomb and two 20mm cannon. The Hs 132C was to have a Heinkel-Hirth 011A engine and be armed with 2,205lb (1,000kg) of bombs and two 20mm cannon.

Henschel Hs 132 data

Role	Single-seat ground-attack aircraft
Ultimate status	Construction
Powerplant	One BMW 003E-2 turbojet, 1,764lb (800kg) st (A), one Jumo 004B turbojet, 1,984lb (900kg) st (B), one HeS 011A turbojet, 2,866lb (1,300kg) st (C)
Maximum speed	485mph (780km/hr) without bomb, 435mph (700km/hr) with bomb at 19,690ft (6,000m) (Hs 132A)
Range	422 miles (680km) (A)
Service ceiling	34,440ft (10,500m) (A)
Endurance	1hr 20min
Weight	7,497lb (3,400kg) loaded (A)
Span	23ft 7½in (7.20m)
Length	29ft 2½in (8.90m)
Wing area	159.3ft^2 (14.8m^2)
Armament	1,102lb (500kg) bomb (A), 1,102lb (500kg) bomb and two MG 151/20 20mm cannon (B), 2,205lb (1,000kg) of bombs and two MG 151/20 20mm cannon (C)

Henschel P.135

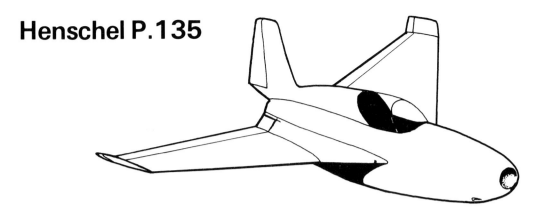

This single-seat mid-wing jet fighter was designed as a private venture in response to the Emergency Fighter specification of 1944 (see FW Ta 183, page 57). It was to be a tailless aircraft with a single central fin. The semi-delta wings were set well back on the fuselage and fitted with dihedral tips. The single HeS 011 turbojet was mounted in the rear fuselage and fed by a nose intake. An armament of four MK 108 cannon was to be fitted, two in the wing roots and two in the lower nose. The scheme got no further than the project stage, being dropped when the OKL favoured one of Focke-Wulf's Ta 183 designs.

Henschel P.135 data

Role	Single-seat jet fighter
Ultimate status	Design
Powerplant	One HeS 011A turbojet, 2,866lb (1,300kg) st
Maximum speed	612 mph at 22,960ft (985km/hr at 7,000m)
Service ceiling	45,930ft (14,000m)
Weight	9,038lb (4,100kg) loaded
Span	30ft 2in (9.20m)
Length	25ft 7in (7.80m)
Wing area	220.6ft² (20.5m²)
Armament	Four MK 108 30mm cannon

Horten Ho IX

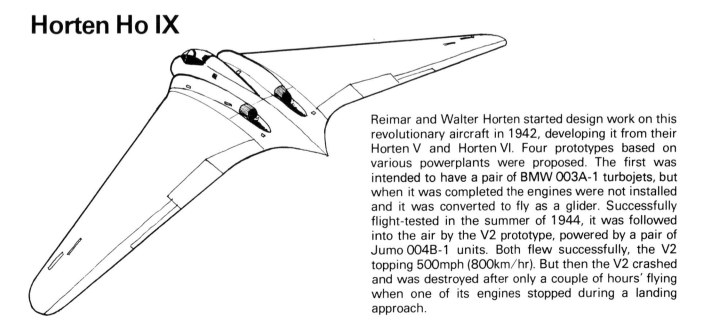

Reimar and Walter Horten started design work on this revolutionary aircraft in 1942, developing it from their Horten V and Horten VI. Four prototypes based on various powerplants were proposed. The first was intended to have a pair of BMW 003A-1 turbojets, but when it was completed the engines were not installed and it was converted to fly as a glider. Successfully flight-tested in the summer of 1944, it was followed into the air by the V2 prototype, powered by a pair of Jumo 004B-1 units. Both flew successfully, the V2 topping 500mph (800km/hr). But then the V2 crashed and was destroyed after only a couple of hours' flying when one of its engines stopped during a landing approach.

In shape the Ho IX was a pure wing with no tail or vertical control surfaces. The centre section was thickened to house a single-seat cockpit and the two engines on either side of it. Two wooden main spars were skinned with plywood and the wingtips and central section were made of metal.

Production versions of the Ho IX were to be single-seat fighter-bombers or night fighters, while the larger Ho IXB two-seat fighter-bomber was to be similarly powered but with a slightly reduced wing span. Length was to be increased to 30ft 2in (9.20m), and a maximum speed of 650mph at 13,120ft (1,045km/hr at 4,000m) was expected, However, before the Hortens could realise these plans, the RLM instructed Gothaer Waggonfabrik to complete the third prototype and to take over all further development of the type under the designation Go 229.

Horten IX data

Role	Single-seat flying-wing jet fighter-bomber testbed
Ultimate status	Flight test
Powerplant	Two Junkers Jumo 004B-1 turbojets, 1,984lb (900kg) st each
Maximum speed	540mph (870km/hr)
Ceiling	52,490ft (16,000km)
Weight	18,740lb (8,500kg) loaded
Span	52ft 6in (16.0m)
Length	24ft 6in (7.47m)

Horten Ho IX V1 (*Heinz Nowarra*)

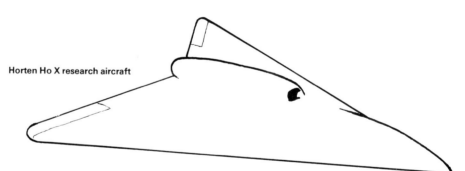

Horten Ho X research aircraft

Horten Ho X

The Horten brothers had long been building all-wing sailplanes, and they turned naturally to jet-powered tailless aircraft during the Second World War. Parts of their Ho X high-speed flying-wing glider were found after the war, though all the drawings and calculations had disappeared. Arrowhead-shaped and weighing 882lb (400kg), it had a very small fin, tricycle undercarriage, nose elevons and wingtip spoilers.

A further development powered by an As 10C pusher piston engine was intended to be followed eventually by an HeS 011-powered research aircraft. Only a sketch of the latter survives, though a post-war Royal Aircraft Establishment report describes it as weighing 13,500-15,000lb (6,120-6,800kg) and having a wing span of 30ft 2½in (9.20m) and a length of 32ft 9½in (10.0m). A very clean design, it had no vertical control surfaces. The wings were sharply swept and the dorsally mounted turbojet exhausted above the wing trailing edge. Maximum speed was estimated at 743mph (1,195km/hr).

Later information refers to a similar aircraft under the same designation but intended to enter service as a fighter in 1946. Like the previous design it was to have a smoothly faired nose and cockpit canopy, and no vertical control surfaces. The turbojet was mounted similarly but with air intakes on either side of the cockpit, rather than the single central intake feeding straight into the engine, of the earlier design. Wing span was significantly increased and the fuselage shortened, and the wings were less sharply swept.

Horten Ho X data

Role	Single-seat flying-wing jet fighter
Ultimate status	Design
Powerplant	One HeS 011A turbojet, 2,866lb (1,300kg) st
Maximum speed	683mph at 19,690ft (1,100km/hr at 6,000m)
Range	1,932 miles (3,110km)
Weight	13,228lb (6,000kg) loaded
Span	45ft 11in (14.0m)
Length	23ft 7½in (7.20m)
Armament	One MK 108 30mm cannon and two MG 131 13mm machine guns

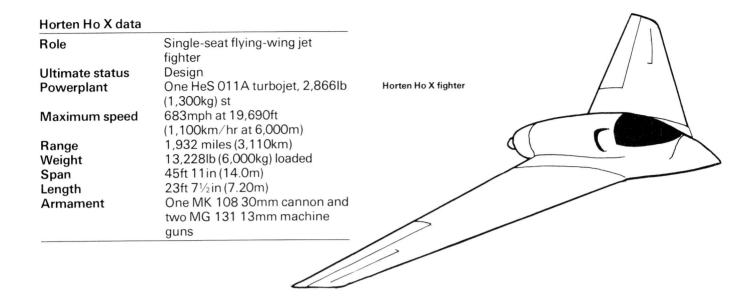

Horten Ho X fighter

Horten Ho XIIIB

The Horten Ho XIII glider was built as a flying testbed to investigate flying-wing control problems. It combined Ho III glider wings with 60° sweepback and a new centre section, and the pilot was to be carried in an underslung nacelle.

Developed from this was a design for a supersonic flying-wing fighter, designated Ho XIIIB. The basic wing shape was retained but the pilot was accommodated in a flush faired cockpit at the base of the large, sharply swept fin. The wings were swept back at 60° and power was to be supplied by a single BMW 003R rocket/turbojet combination. This unit could deliver a total thrust of over 3,000lb (1,360kg) for short periods of emergency high-speed action.

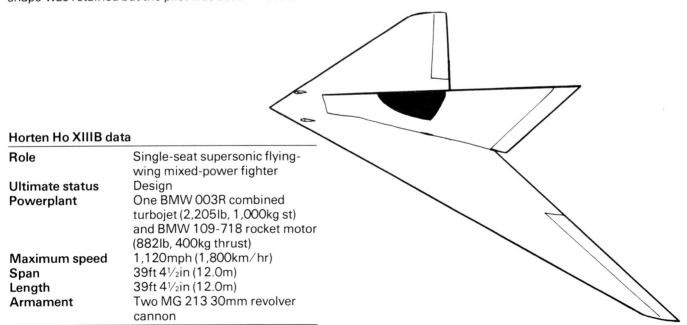

Horten Ho XIIIB data

Role	Single-seat supersonic flying-wing mixed-power fighter
Ultimate status	Design
Powerplant	One BMW 003R combined turbojet (2,205lb, 1,000kg st) and BMW 109-718 rocket motor (882lb, 400kg thrust)
Maximum speed	1,120mph (1,800km/hr)
Span	39ft 4½in (12.0m)
Length	39ft 4½in (12.0m)
Armament	Two MG 213 30mm revolver cannon

Horten Ho XVIII

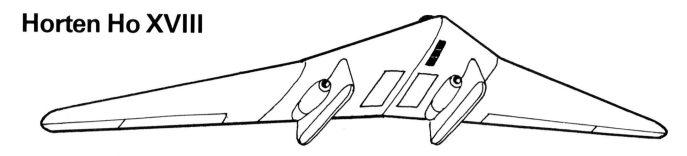

Horten Ho XVIIIA

Late in the war the Hortens produced two sketch designs of flying wings in response to an RLM specification for a fast bomber with a range of nearly 5,000 miles (8,000km), payload of 8,000lb (3,600kg) and minimum cruising speed of 500mph (800km/hr). Service-entry date was to be 1946-47. A maximum landing speed of 100mph (160km/h) and a 3,000ft (900m) take-off distance with rocket-assisted take-off gear were also specified.

The Ho XVIIIA had an arrowhead planform, with no vertical control surfaces. A bubble canopy covered the pilot's station and all four crew were enclosed in a pressure cabin. Four HeS 011A turbojets were mounted in pairs on either side of the fixed, eight-wheeled undercarriage legs, the bogies of which were faired over during flight. The outer wings were to be made of wood and with a shallow dihedral, the centre section of wood or steel and light alloy. The main and auxiliary spars were of box section.

The significantly larger Ho XVIIIB had a large central fin and rudder, with the crew accommodated at its base behind a glazed leading edge. Located at the base of the fin trailing edge was a twin-cannon barbette which was operated remotely and sighted by periscope; two more cannon were fixed in the nose. Power was to be supplied by six Jumo 004H turbojets slung beneath the wing in a single housing.

The Ho XVIII design was prepared to the same specification that resulted in the Junkers EF.130 (see page 99) and Messerschmitt P.1107 (see page 129) jet bomber projects.

Horten Ho XVIII data

Role	Four/six-seat long-range jet bomber
Ultimate status	Design
Powerplant	Four HeS 011A turbojets (2,866lb, 1,300kg st each) (A), six Jumo 004H turbojets (2,426lb 1,100kg st each) (B)
Maximum speed	534mph at 22,960ft (860km/hr at 7,000m) (A), 560mph at 19,680ft (900km/hr at 6,000m) (B)
Range	3,356 miles (5,400km) (A), 5,593 miles (9,000km) (B)
Weight	73,000lb (33,100kg) (A), 97,000lb (44,000kg) (B) (both loaded)
Span	98ft 6in (30.0m) (A), 137ft 9½in (42.0m) (B)
Length	62ft 4in (19.0m) (B)
Armament	Four MG 213 remotely controlled 30mm revolver cannon and 8,818lb (4,000kg) of bombs

Horten Ho XVIIIB

Junkers EF.008

Junkers designed a large number of jet-powered aircraft bearing the "EF" designation (*Entwicklungsflugzeug*, development aircraft). Most of them, like the EF.008, were abandoned at an early stage, and very few details remain.

The 008 was designed as a two-seat bomber powered by four turbojets slung individually beneath the straight shoulder wings. A single 20mm cannon was located beneath the pressurised cabin in the nose.

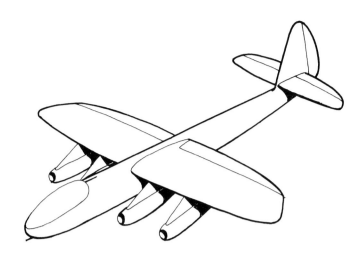

Junkers EF.009

The design of this very early jet fighter project was centred upon overcoming the problems created by the limited power available from the first generation of turbojets. It was to be powered by no fewer than ten small turbojets, each about two feet long and spaced around the circumference of the nose section. The rapid advance in jet technology made the use of so many engines unnecessary, however, and the project was dropped.

The wings were set low and were of low aspect ratio with slightly swept leading edges. Landing gear was to comprise a central retractable skid and a small fixed tailskid.

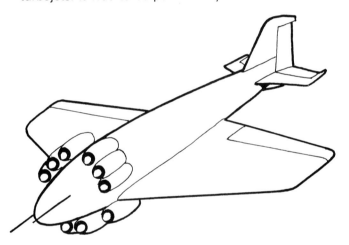

Junkers EF.009 data

Role	Single-seat jet fighter
Ultimate status	Design
Powerplant	Ten small turbojets
Maximum speed	540-590mph (870-950km/hr)
Weight	4,410lb (2,000kg) loaded
Span	13ft 1½in (4.0m)
Length	16ft 5in (5.0m)
Armament	Two 30mm cannon

Junkers EF.010

An early twin-jet design, this project was described simply as a "record aircraft" and resembled conventional single piston-engined aircraft of the time except that it had a long pointed nose where the engine would normally have been. The two short turbojets were mounted on long pylons fixed to the lower wing leading edges. So far forward were the engines that even the rear of each pod was some distance in front of the wings. The undercarriage was of the tailwheel type, which was later found unsuitable for jet aircraft.

The design did not go into development, presumably because record-breaking was considered too frivolous at a time when military aircraft were in great demand.

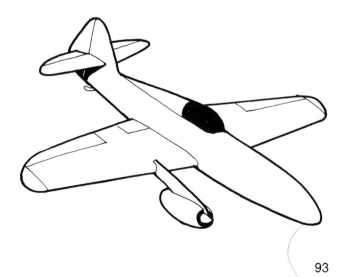

Junkers EF.011

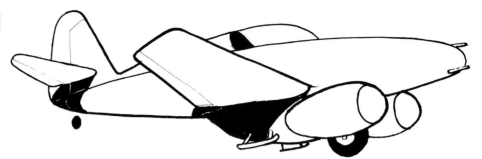

Very similar in layout to the EF.010, the EF.011 also had straight wings, but with blunt tips, and a shortened nose. Designed as a single-seat fighter, it was to be powered by a pair of Walter rocket motors fixed beneath the fuselage on either side of a large central pod housing a single central landing wheel and two outrigger stabilising skids. Armament was to be two 20mm cannon mounted in the nose.

Junkers EF.012

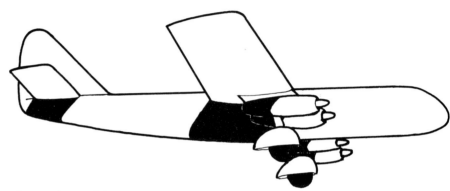

This project got as far as wind-tunnel tests in model form, but it went no further and soon vanished amongst the host of other early Junkers jet aircraft designs. The 012 was a straight mid-wing bomber powered by four turbojets mounted in pairs suspended on pylons beneath each wing leading edge. It was to have a crew of two and a tailwheel-type undercarriage with a fixed main leg located between the units of each engine pair. The wheels were spatted, like those of the Ju 87, in an attempt to reduce drag. Wing span was 50ft 10½in (15.50m) and an armament of two remotely controlled turrets with a single 20mm cannon each was envisaged.

Junkers EF.015

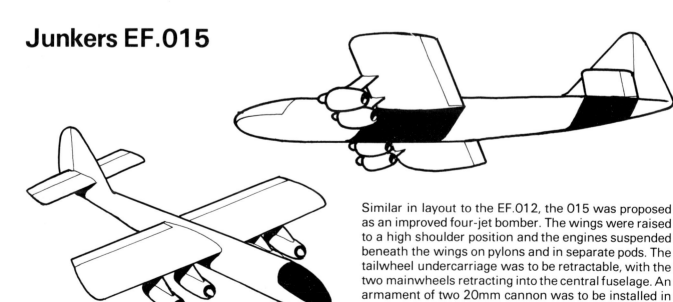

Similar in layout to the EF.012, the 015 was proposed as an improved four-jet bomber. The wings were raised to a high shoulder position and the engines suspended beneath the wings on pylons and in separate pods. The tailwheel undercarriage was to be retractable, with the two mainwheels retracting into the central fuselage. An armament of two 20mm cannon was to be installed in the nose.

Junkers EF.017

This jet fighter project bore a very strong resemblance to the He 280 (see page 78), especially in its wing planform and engine location. The tailplane was more conventional, however, with a single central fin, and the undercarriage was of the tailwheel type. Proposed armament consisted of two 20mm cannon in the nose.

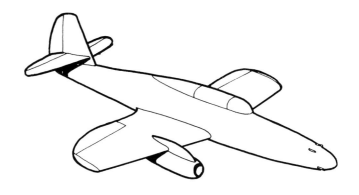

Junkers EF.018

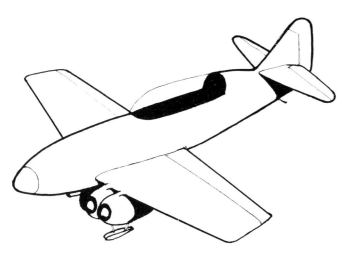

Another early fighter project from the Junkers drawing boards, the EF.018 was powered by no fewer than four short turbojets positioned in pairs beneath each wing leading edge. A low straight-wing aircraft, it had a large cockpit canopy, conventional tailplane and an undercarriage comprising a single central mainwheel, two stabilising skids located with each engine pair, and a tailskid. Armament was to be two 20mm cannon located in the lower nose or wing leading edges.

Junkers EF.019

The EF.019 bore a strong resemblance to the later British Gloster Meteor fighter, having straight low wings which seemed to pass through the centre of the turbojets mounted at one-third span. It was intended for fighter duties and had a conventional tailplane and tailwheel undercarriage. Two 20mm and two 30mm cannon were to be mounted in the nose.

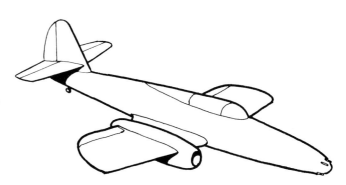

Junkers EF.112

An early twin-jet bomber project, the EF.112 had a straight mid-wing with slightly swept leading edges. No technical information survives but a model is known to have been built for wind-tunnel testing.

Junkers EF.116

Described simply as a "jet bomber", the EF.116 had a long, cylindrical fuselage and mid-set wings which were swept back for most of their length and forward for the last quarter of the span. As with the EF.112, very little technical information survives – number and disposition of engines is unknown, for instance – but a wind-tunnel model was produced before the project was finally dropped.

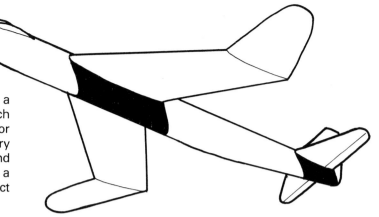

Junkers EF.122

One of the many developments of the Ju 287, this bomber design was built in wind-tunnel model form only. Like the Ju 287, it had forward-swept, mid-set wings, with a single turbojet mounted above each at about one-third span. Two more turbojets were to be fixed on either side of the forward fuselage.

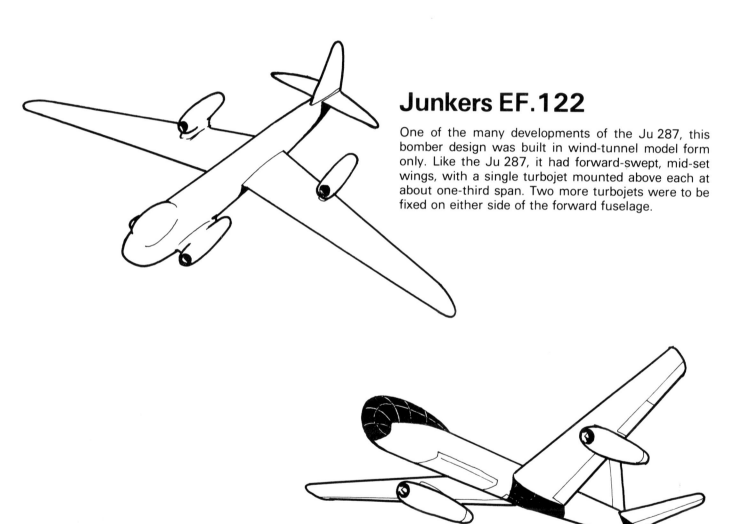

Junkers EF.125

The EF.125 was the final proposed development of the Ju 287 jet bomber design. It was to have forward-swept wings set low on the fuselage and each carrying one underslung turbojet, either the 6,614lb (3,000kg) st Jumo 012 or the 7,496lb (3,400kg) st BMW 018. The engines were located well forward on each wing in order to minimise the twisting to which forward-swept wings were prone. Similar in size to the Ju 287, the EF.125 had an estimated maximum speed of over 680mph (1,100km/hr).

Junkers EF.126 Elli

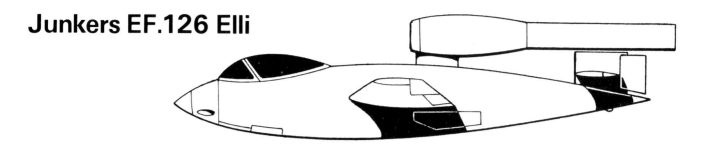

Submitted as an entrant in the RLM's *Miniaturjäger* competition late in 1944, the EF.126 *Elli* project found itself pitted against designs from Blohm und Voss and Heinkel. The Junkers design was to have an all-metal fuselage with a mid-set wooden wing. Layout was very similar to that of the Fi 103 R-IV *Reichenberg*, with the cockpit positioned in the nose, and the powerplant was to be a single Argus pulsejet. The specification, issued in November 1944, called for a very simple jet fighter which could be built more rapidly than the *Volksjäger* (He 162) but which was not semi-expendable like the Ba 349 (see page 24).

The initial design featured a landing skid, but the production version was to have a simple nosewheel undercarriage. Provision was made for a bomb load of 882lb (400kg). The latter demand foresaw the possibility that the project might not be accepted as a fighter – probably as a result of the significant loss of efficiency suffered by pulsejets at increasing heights – and so would have to be pressed into service as a ground attacker. Such was the relatively low power of the 1,102lb (500kg) thrust Argus 109-044 pulsejet, take-off was to be assisted by two solid-fuel rockets or a winch.

The EF.126 did not fly during the war, though a full-scale mock-up was built. After the war the Russians built an unpowered prototype which eventually crash-landed, killing the pilot.

Junkers EF.126 data

Role	Single-seat jet fighter or ground attack aircraft
Ultimate status	Unpowered flight test
Powerplant	One Argus 109-044 pulsejet, 1,100 2lb (500kg) thrust
Maximum speed	485mph (780km/hr) at sea level, 290mph (470km/hr) at 60 per cent thrust, 422mph (1,680km/hr) with external bomb load
Range	183 miles (295km) (23min) at maximum speed or 218 miles (350km) (45min) at 60 per cent thrust
Weight	2,462lb (1,100kg) empty, 6,172lb (2,800kg) loaded
Span	20ft 9in (6.32m)
Length	24ft 11¼in (7.60m)
Wing area	95.8ft^2 (8.9m^2)
Armament	Two MG 151/20 20mm cannon, plus 882lb (400kg) of bombs or 12 *Panzerblitz* rockets for ground attack

Junkers EF.127 Walli

Designed as a target defence interceptor, the EF.127 was based on the EF.126, with a wooden mid-set wing and all-metal fuselage. Code-named *Walli*, the EF.127 differed from the EF.126 in having a fuselage-mounted Walter HWK 109-509A-2 rocket motor as its main powerplant. It was to be used in the same role as the Me 163 and Bachem *Natter*, but the inadequate endurance of the rocket motors led to the suspension of work on the *Walli* and a parallel development, the Heinkel *Julia*, resumption of the programme being dependent

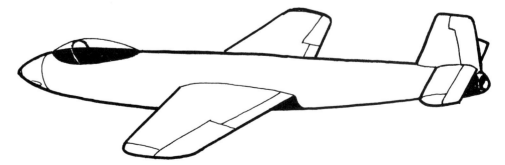

upon results obtained with the Me 163, 263 and 262. It was originally expected to have the *Walli* ready by August 1945, and one was discovered to be under construction when the war ended.

A wheeled undercarriage is indicated in drawings, but the landing was expected to be made on skids. Two assisted take-off rocket units of 2,205lb (1,000kg) thrust each were to be fitted.

Junkers EF.127 data

Role	Rocket-powered single-seat target defence interceptor
Ultimate status	Construction
Powerplant	One Walter HWK 109-509A-2 rocket motor (3,307lb, 1,500kg thrust) and two assisted take-off rockets (2,205lb, 1,000kg thrust each)
Maximum speed	630mph (1,015km/hr) at sea level, 590mph at 19,690ft (950km/hr at 6,000m), 560mph at 36,090ft (900km/hr at 11,000m)
Range	66 miles (105km) at 435mph (700km/hr) after climbing to 16,400ft (5,000m)
Climb rate	26,250ft/min (8,000m/min)
Weight	6,140lb (2,785kg) loaded
Span	20ft 6½in (6.26m)
Length	24ft 5½in (7.45m)
Wing area	95.7ft² (8.83m²)
Armament	Two MK 108 30mm cannon, 60 rounds per gun

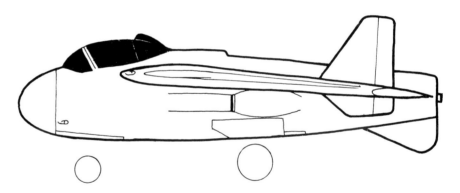

Junkers EF.128

Submitted for the Emergency Fighter competiton, the EF.128 was a tailless aircraft powered by a single HeS 011 turbojet. The all-wooden shoulder-mounted wings were swept back 45°, with the engine air intakes located below them on either side of the fuselage. The vertical control surfaces were positioned at half-span on each wing trailing edge. The pilot was protected against 0.50in (12.7mm) fire from ahead and 20mm shells from behind.

Good results were obtained with a completed wind-tunnel model, and a mock-up fuselage with an HeS 011 turbojet was built for tests in which it was to be mounted above a Ju 88 fuselage. Though promising, the project was ultimately rejected in favour of the Focke-Wulf Ta 183 *Projekt* I.

A further variation with a lengthened fuselage was planned as a two-seat all-weather fighter.

Junkers EF.128 data

Role	Single-seat jet fighter
Ultimate status	Wind-tunnel tests
Powerplant	One HeS 011A turbojet, 2,866lb (1,300kg)
Maximum speed	547mph (880km/hr) at sea level, 590mph at 19,690ft (950km/hr at 6,000m)
Range	1,120 miles (1,800km)
Ceiling	45,000ft (13,700m)
Span	29ft 3in (8.90m)
Length	22ft 11½in (7.0m)
Armament	Two MK 108 30mm cannon, 100 rounds per gun (two additional MK 108 optional)

Junkers EF.130

Several different designs were produced under the EF.130 designation, and this has given rise to some confusion, particularly concerning the positioning of the four jet engines. A jet bomber of flying-wing layout, the EF.130 was projected early in 1945 at the same time as the Horten XVIII and Messerschmitt P.1107 (see pages 92 and 129). Fins were mounted on the wing trailing edges at mid-span, with flaps fitted inboard and ailerons outboard. The wings were to be of all-wooden construction, with a metal central fuselage section. Powerplant was to be either four HeS 011A turbojets or four BMW 003Es podded singly and mounted on short pylons in a row along the rear of the wing centre section. A retractable tricycle undercarriage was to be fitted with fuel tanks located in the centre section and wings. The project was not developed.

Junkers EF.130 data

Role	Flying-wing jet bomber
Ultimate status	Design
Powerplant	Four HeS 011A turbojets (2,866lb, 1,300kg st each) or four BMW 003E turbojets (1,764lb, 800kg st each)
Maximum speed	615mph at 19,690 (990km/hr at 6,000m) (with HeS 011A)
Range	3,666 miles (5,900km)
Weight	83,795lb (38,000kg) loaded
Span	78ft 9in (24.0m)
Wing area	1,290ft^2 (119.8m^2)
Armament	6,500lb (2,950kg) of bombs

Junkers EF.132

This very poorly documented high-altitude bomber project was designed around a pressurised cabin and six turbojets.

Junkers EF.135

The airframe of this aircraft seems to have been similar to that of the EF.130, but with two tailbooms. The nature of the powerplant is not known, but one reference suggests that it was of the mixed turbojet/piston type.

Junkers Ju 248

Ju 248 was the designation given to a development of the Messerschmitt Me 163 rocket fighter, the Me 163D, which had been handed over for completion to the Junkers design staff. By the time a first prototype had been built by Junkers in August 1944 it had been redesignated once again with a new Messerschmitt identity, Me 263, and it was under this guise that the aircraft continued its development. See Messerschmitt Me 263 (page 120) for further information.

Junkers Ju 267

This aircraft was described soon after the war as an experimental type with turbojet engines mounted in exterior nacelles.

Junkers Ju 268 (Arado E.377A) Mistel 5

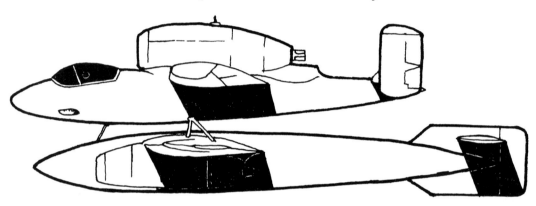

Ju 268 was the Junkers designation given to an unusual composite flying-bomb design which was also known as *Mistel* 5 (*Mistel*, Mistletoe, was the standard code name for German composite aircraft projects). The upper component of the composite was to be a Heinkel He 162 single-jet fighter and it was to be connected to a large twin-jet flying bomb by a number of explosive bolts. Originally the lower component was an Arado project, known as the E.377 in glider form and E.377A when powered by two BMW 003 turbojets.

Three alternative warheads were projected. The first consisted of a single 4,410lb (2,000kg) bomb cradled in the lower forward section of the fuselage. The second was a steel-cased 7,700lb (3,490kg) hollow charge positioned near the aircraft's centre of gravity. The third was a solid-charge head of similar weight and located in the same position.

Junkers designated the unmanned aircraft the 8-268. It was to be of wooden construction, with straight shoulder wings under which were slung two BMW 003 turbojets. The composite was to accelerate down the runway on a Rheinmetall-Borsig trolley which would be jettisoned after take-off.

It was also planned to investigate the use of the 8-268 as a piloted aircraft with conventional controls.

Junkers Ju 268 data

Role	Fighter-guided unmanned jet flying bomb
Ultimate status	Design
Powerplant	Two BMW 003A turbojets (8-268), 1,764lb (800kg) st each
Maximum speed	485mph (780km/hr) at sea level and 516mph at 19,690ft (830km/hr at 6,000m) in composite form
Weight	9,447lb (4,285kg) (8-268), 3,803lb (1,725kg) (He 162), 13,250lb (6,010kg) total all empty 23,104lb (10,480kg) (8-268), 6,823lb (3,095kg) (He 162), 29,927lb (13,575kg) total all loaded
Span	37ft 8½in (11.50m) (8-268)
Wing area	237.0ft² (22.0m²) (8-268), 120.56ft² (11.2m²) (He 162)

Junkers Ju 287

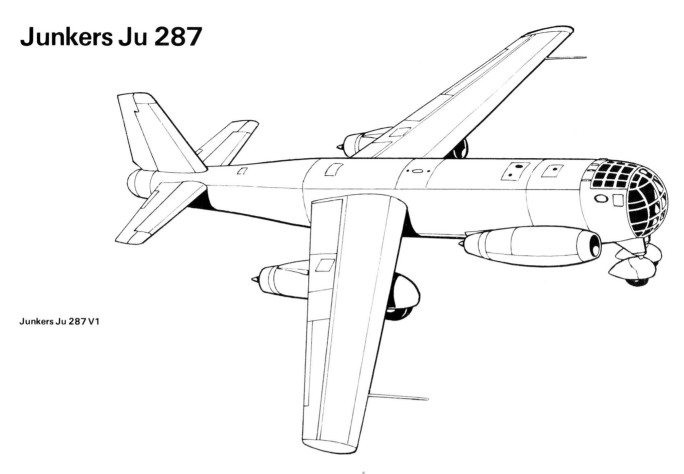

Junkers Ju 287 V1

In 1943 the *Technisches Amt* (Technical Department) of the RLM decided to specify jet propulsion for future fighters and bombers in an effort to regain air superiority. One jet bomber, the Ar 234 *Blitz*, was already flying at this time, and though for its small size it was a reasonably effective aircraft, its straight wings kept its top speed below the theoretical maximum. Attention was therefore turned to a Junkers project that had been started in June 1943. The aircraft was to have swept-back wings and four turbojets, one beneath each wing and one either side of the forward fuselage. But the low-speed stability problems of swept wings had not at that time been solved, so in order to retain the high-speed potential of sweep while at the same time avoiding low-speed stalls, the designers compromised by giving the wings 25° of forward sweep. The wings were attached to the fuselage at about its midpoint, giving the aircraft the appearance of having an exaggeratedly long nose.

The RLM ordered the first prototype of the Ju 287 in March 1944. It was based on the fuselage of a He 177 and, as the initial trials were to be conducted at low speed, the undercarriage was fixed. The first flight took place on August 16, 1944, and though the aircraft flew well at low speed there were some indications of wing twisting. This phenomenon was peculiar to the forward-swept wings and was characterised by an upward twisting of the wingtips and leading edges in a fast climb. The resulting effective change in the aerofoil section would cause the aircraft to nose up still more, exacerbating the problem. *(continued on page 103)*

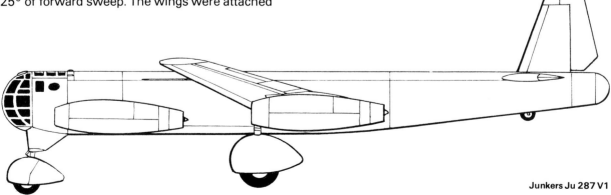

Junkers Ju 287 V1

Above: Ju 287 V1 fitted with Walter rocket-assisted take-off packs (*IWM*)

Right: Ju 287 V1 during its flight-test programme (*via Pilot Press*)

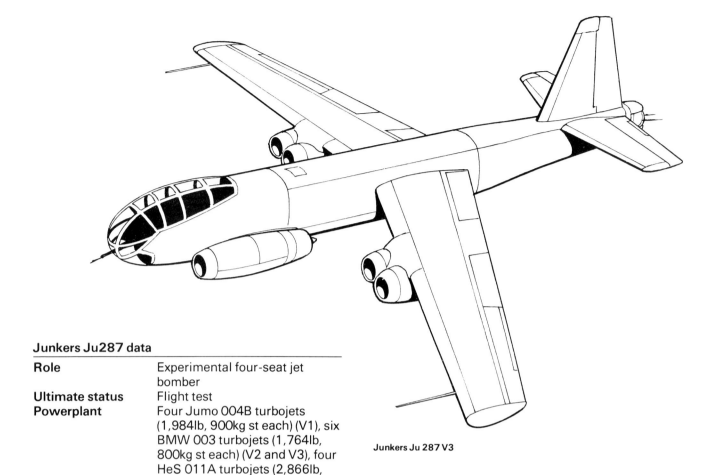

Junkers Ju 287 V3

Junkers Ju287 data

Role	Experimental four-seat jet bomber
Ultimate status	Flight test
Powerplant	Four Jumo 004B turbojets (1,984lb, 900kg st each) (V1), six BMW 003 turbojets (1,764lb, 800kg st each) (V2 and V3), four HeS 011A turbojets (2,866lb, 1,300kg st each) (production examples)
Maximum speed	348mph at 20,000ft (560km/hr at 6,100m), 398mph (640km/hr) in a dive (V1) 621mph (1,000km/hr) with swept-back wings (1948, V2) 537mph at 16,400ft (865km/hr at 5,000m) (V3) 547mph at 20,000ft (880km/hr at 6,100m) (production examples)
Range	932 miles (1,500km) (V1), 2,760 miles (4,440km) (production)
Service ceiling	35,400ft (10,800m)
Weight	22,553lb (10,230kg) empty, 44,092lb (20,000kg) loaded (V1), 68,500lb (31,070kg) loaded (production)
Span	65ft 11¾in (20.10m) (V1), 63ft 6in (19.35m) (production)
Length	60ft (18.60m) (V1), 64ft (19.50m) (production)
Wing area	656.6ft² (61.0m²) (V1)
Armament	Two MG 131 13mm machine guns and 8,808lb (4,000kg) of bombs

Meanwhile, a six-engined second prototype (V2) was being built for high-speed research. The engines were mounted in clusters of three, one under each wing, in such a way as to counterbalance the wing-twisting moment.

In July 1944 all German bomber production was abandoned in favour of fighters, but Junkers slowly continued building the V2 and had already started work on a third prototype (V3) with relocated engines. The company's reward was a sudden production order in March 1945, at which point it was expected to have about a hundred aircraft completed by the end of September 1945. It was intended to power initial production aircraft with four HeS 011s, and other versions with six BMW 003s positioned as on the V3. There was to be a pressure cabin for the four-man crew.

When the war ended the Russians took the first two prototypes and a group of German engineers back to the Soviet Union, where tests continued until 1948. The V2 was fitted with swept-back wings and is reported to have attained a speed of 621mph (1,000km/hr).

Lippisch-Sander Ente

Although in 1928 Fritz von Opel had started work on his own Rak-1 aeroplane and Rak-2 rocket car, he also proposed that Alexander Lippisch, who at that time was just beginning his career as an aircraft designer, should design a rocket-powered aircraft as part of von Opel's publicity and fund-raising project (see Espenlaub Rak-3, page 51). The result was a canard (tail-first) glider with fins mounted on the mainplanes, which were themselves mounted above the rear fuselage. The horizontal control surfaces were fitted above the forward fuselage. The underslung fuselage housed the pilot and a pair of 44lb-thrust Sander solid-fuel rockets which were to be fired in sequence, each having a duration of 30sec.

Although work on the *Ente* (Duck) was started after the inception of the Rak-1, it was the first to be ready for rocket flight. It took to the air for the first time on June 11, 1928, three weeks after von Opel's Rak-2 rocket car had succeeded in attaining a speed of 143mph (230km/hr).

The *Ente* covered approximately three-quarters of a mile in 70sec under rocket power, reaching a speed of around 60mph (100km/hr), on that first flight. This cannot however be considered as the first true and complete piloted rocket aircraft flight, as the aircraft had to be assisted into the air by a rubber cord which catapulted it along a wooden track (see Opel-Hatry Rak-1, page 133). Friedrich Stamer, the pilot, reported that the aircraft was very difficult to control, and Lippisch abandoned the project after only the third flight when one of the rockets exploded in flight.

Lippisch DM-1

In 1937 Doctor Alexander Lippisch assumed the leadership of a design team developing the RLM's *Projekt* X, which was eventually to become the Me 163 rocket-powered interceptor. Five years later he left the project, just as the first *Komet* production prototypes were being completed, to lead another research team consisting of students of aircraft construction from Darmstadt and Munich universities. Working with the help of the DFS on a programme intended to lead to the development of a fast interceptor, Lippisch and the students produced a series of revolutionary aircraft, designated with a DM prefix in recognition of the two universities.

The first design to come from the team was the DM-1, which was found in an incomplete state at the end of the war by the Americans. Construction work had started on the DM-1 flying testbed in November 1944. A pure delta with 60°-swept leading edges, it was to be used initially as a glider to investigate flight characteristics. Fin and rudder shape mirrored that of the wings, and the pilot was accommodated in a cockpit at the base of the fin.

It was originally intended to carry the DM-1 on the back of a Siebel Si 204 to a height of 25,900ft (7,900m), from which it would dive to an anticipated speed of 348mph (560km/hr). At a later stage the DM-1 was to be flown at a speed of 497mph (800km/hr) under the power of a rocket motor. At the other end of the speed range, the aerodynamic characteristics of this little single-seat aircraft were such that a landing speed of only 44mph (70km/hr) was expected.

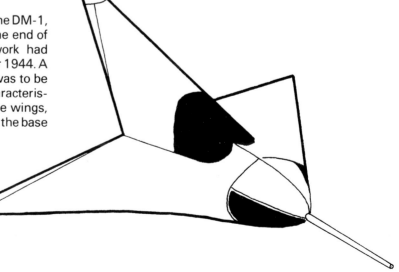

The Americans shipped the prototype back to the USA for completion and flight testing, and the resulting data were incorporated into the design of the many US delta-wing aircraft which appeared in subsequent years.

Lippisch DM-1 data

Role	High-speed delta-wing research glider
Ultimate status	Flight test
Powerplant	Rocket motor (never fitted)
Maximum speed	348mph (560km/hr) in unpowered dive (497mph, 800km/hr estimated with rocket power)
Weight	655lb (297kg) empty, 1,015lb (460kg) loaded
Span	19ft 5in (5.92m)
Length	21ft 8in (6.60m)
Wing area	215.3ft² (20.0m²)

Lippisch DM-2, DM-3 and DM-4

Lippisch and his team of students designed an aircraft based on the DM-1 and powered by a single turbojet. Designated the DM-2, it was to be used to investigate the effects of high speeds on delta airframes. A speed of 497-746mph (800-1,200km/hr) at high altitude was expected, but the DM-2 never got off the drawing board.

An even faster version of the delta-wing DM-1 and DM-2, powered by two rocket motors and designated DM-3, was expected to achieve a speed of over 1,240mph (2,000km/hr) at high altitude. This design was not built and nor was the DM-4, intended as a supersonic delta-wing research aircraft fitted with a variety of powerplants.

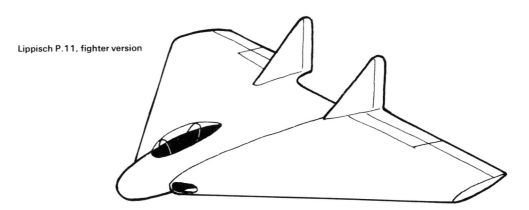

Lippisch P.11, fighter version

Lippisch P.11

After his time with the Darmstadt and Munich universities team Lippisch turned his attentions in 1944 to more warlike aircraft, whereupon he seems to have once more received the backing of the RLM.

Dating from 1944, the P.11 appears to have been designed in at least two forms. The first to come to light was a flying-wing fighter with a short nose, bubble canopy, and wing leading edges swept back at less than 45°. The trailing edges were straight, apart from a cut-out in the centre section, and the wingtips were cut off square. A pair of triangular fins were located on either side of the centre-section cut-out, and two Jumo 004 turbojets exhausted below and between them. An air intake was positioned in each wing leading edge on either side of the nose section. Fuel was to be carried in large tanks in each wing.

Lippisch P.11, fighter version

This version was to be developed in parallel with the Ho 229 (Go 229) flying-wing fighter, according to the minutes of a meeting of German aircraft development programme planners on November 21 and 22, 1944. It was proposed that the Lippisch P.11 fighter would be further developed in collaboration with the Henschel aircraft company.

A second version was similarly powered but was slightly larger, with a wing span of 41ft 4in (12.60m) and a length of 26ft 6in (8.07m). The cockpit canopy was removed from the upper surface and the pilot was located in the extreme nose behind a flush-fitting canopy. It was described as being designed for high-speed bombing duties and based on the tailless Me 265 bomber project, with estimated maximum speed given as 560 mph at 32,800ft (900km/hr at 10,000m).

Lippisch P.11 data

Role	Flying-wing jet fighter or bomber (data for fighter version)
Ultimate status	Design
Powerplant	Two Junkers Jumo 004B turbojets, 1,984lb (900kg) st each
Maximum speed	645mph at 19,350ft (1,200km/hr at 5,900m), 528mph (850km/hr) cruising speed
Range	1,864 miles (3,000km)
Weight	16,005lb (7,260kg) all-up
Span	35ft 5in (10.80m)
Length	23ft (7.0m)
Wing area	538.2ft^2 (50.0m^2)
Armament	Two MK 103 30mm cannon

Lippisch P.12

This unusual little flying-wing project was to be powered by a single liquid-fuel ramjet. A large oval air intake in the nose fed the powerplant, which was located between the delta wings. These were turned downward at the tips to form stabilisers for the single central landing wheel. The pilot was seated in a finely blended cabin located above the ramjet combustion chamber, the cockpit canopy forming the base of the single triangular fin. Wing area was approximately 130ft^2 (12.0m^2) and aspect ratio 1.33. This project was abandoned in favour of the P.13.

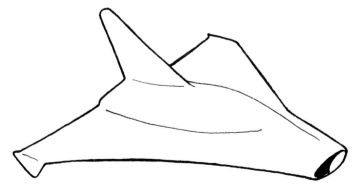

Lippisch P.13

Strongly reminiscent of the DM-1 high-speed research glider, the P.13 was originally designed as a two-seater but soon assumed the form illustrated here and designated P.13a.

The RLM first began to show interest in ramjet propulsion in November 1944, when it was decided to co-ordinate research with a view to developing a ramjet-powered fighter. One particularly novel aspect of the resulting specification was a request that provision be made for the "foam coal" fuel being developed by Professor Lippisch at that time.

The P.13a was based on the DM series of high-speed delta-wing research aircraft, and it was to be powered by a single coal-fuelled ramjet. It was originally proposed that the solid fuel for the ramjet was to take the form of small pieces of brown coal carried in a wire-mesh container set in the duct at a shallow angle to the airstream. The flow of air through the lower portion of the duct was therefore obstructed, and it was calculated that the burning of the fuel in the air passing through the mesh container would produce carbon monoxide, which would then mix with oxygen in the air passing through the unobstructed upper portion of the duct to form high-pressure carbon-dioxide. But this arrangement was soon dropped in favour of a design which incorporated a circular basket which was supported inside the duct and rotated on a vertical axis at 60rpm. Combustion was to be started by a gas flame, with assistance from liquid fuel. Coal granules were proposed as an alternative, more easily combustible material.

Wind-tunnel tests showed that the extremely clean P.13 was stable up to a speed of Mach 2.5. The delta wings of the P.13a were swept back at 60° and the pilot was housed in the very large triangular fin, sitting above the ramjet combustion chamber. A rocket motor was to be used for take-off and acceleration up to ramjet operating speed at around 200mph (320km/hr), at which point the powdered-coal fuel was ignited by an oil burner. It was estimated that 1,764lb (800kg) of coal would give an endurance of 45min.

By the end of the war various companies were designing four or five ram-jet-powered fighters – all except the P.13 with liquid-fuel engines – but none was completed.

Lippisch P.13 data

Role	Ramjet/rocket fighter
Ultimate status	Design
Powerplant	One Lippisch coal-burning ramjet plus auxiliary rocket motor
Maximum speed	1,025mph (1,650km/hr) at high altitude
Endurance	45min on 1,764lb (800kg coal fuel)
Weight	5,060lb (2,295kg) loaded
Span	19ft 5in (5.92m)
Length	22ft (6.70m)
Wing area	215ft^2 (20.0m^2)

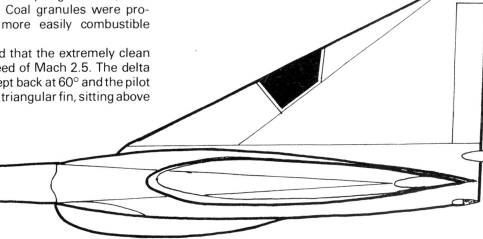

Lippisch P.14

In designing the P.14 Lippisch moved away from the flying-wing arrangement which had characterised his earlier projects. Only the barest of details have survived, but the P.14 is reported to have had a constant-chord but otherwise conventional wing and tail. A number of coal-fuelled ramjets were to be built into the wings.

Lippisch P.15

A development of the Me 163, the P.15 was to be powered by a single centrally mounted HeS 011A turbojet. The nose of the aircraft, housing two 30mm cannon, was to be taken from the He 162. The wings, with a cannon in each root, were from the Me 163, their span being increased by the location of the air intakes in each leading-edge wing root. Other borrowed parts included the Bf 109 main undercarriage.

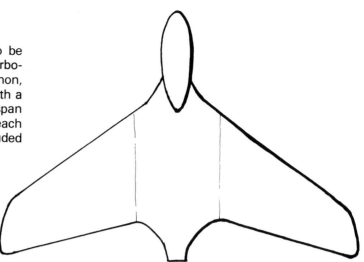

Lippisch P.20 (Messerschmitt P.20)

The P.20 was a further development of the Me 163 target-defence interceptor, in this case powered by a single Jumo 004 turbojet. An air intake set low in the extreme nose was to feed the fuselage-mounted engine, but otherwise the aircraft was very similar in appearance to the Me 163 *Komet*, with comparable wing span but slightly greater length.

This project, which never left the drawing board, has also been referred to as the Messerschmitt P.20.

Lippisch P.20 data

Role	Single-seat jet fighter
Ultimate status	Design
Powerplant	One Junkers Jumo 004C turbojet, 2,205lb (1,000kg) st
Maximum speed	568mph at 26,250ft (915km/hr at 8,000m)
Range	350 miles (560km)
Service ceiling	40,030ft (12,200m)
Weight	7,496lb (3,400kg) loaded
Span	30ft 6in (9.30m)
Length	20ft (6.10m)
Armament	Four MK 108 30mm cannon

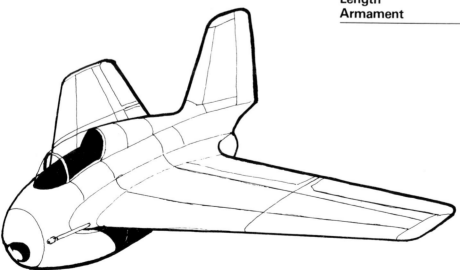

Lippisch supersonic flying wing

One of the most adventurous of all the visionary designs to come from Alexander Lippisch was a flying wing powered by a single ramjet which was intended to achieve twice the speed of sound. The delta wings were to have even more sweepback than those of the P.13, while that type's large triangular fin was replaced with a very much smaller unit above the central wing trailing edge. The pilot was located beneath a very shallow canopy and probably assumed a prone or near-prone position.

Lippisch P.01-111

During the *Projekt* X period (see Me 163, page 113) Lippisch produced a number of designs for small tailless fighters. Most remained little more than sketches but the trend was steadily towards the final Me 163 *Komet* shape.

In October 1939 the P.01-111, clearly based on the tailless DFS 194, was drawn up as a fighter. Although the powerplant is not known, the location of an air intake in the nose indicates that it was to be a ramjet or turbojet. The wings had slight dihedral and leading edges swept back at 30°, while the trailing edges were nearly straight and the tips were blunted.

Lippisch P.01-111 data

Role	Tailless jet or ramjet fighter
Ultimate status	Design
Span	25ft 2½in (7.68m)
Length	21ft 7in (6.57m)
Armament	Two 20mm cannon

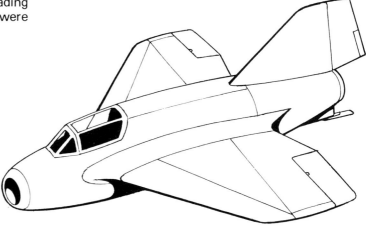

Lippisch P.01-113

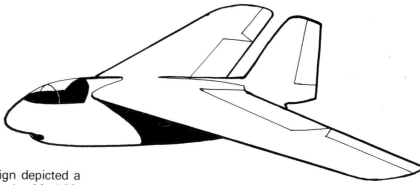

Dated July 7, 1940, the P.01-113 design depicted a shoulder-wing tailless fighter similar to the Me 163. The dual powerplant was to comprise a 1,764lb (800kg) st BMW 003 turbojet, fed by an air intake beneath the aircraft's nose, and a BMW 3304 rocket motor. The wing leading edges were swept back at 30° and the pilot was located behind a smoothly faired canopy in the nose.

Lippisch P.01-113 data

Role	Tailless jet/rocket fighter
Ultimate status	Design
Span	29ft 6½in (8.90m)
Length	24ft 7in (7.50m)
Armament	Two 30mm cannon

Lippisch P.01-114

Two designs seem to have been produced under the P.01-114 designation. The first, dated July 19, 1940, was an experimental unarmed aircraft with a shoulder wing and no tailplane. The aircraft had the appearance of a truncated modern sailplane with 30°-swept wings and was to be powered by a variable-thrust BMW 109-510 rocket motor producing between 660lb and 3,307lb (300-1,500kg) of thrust. Wing span was to be 29ft 6½in (8.90m) and length 20ft 7½in (6.28m).

The second version was tailless and bore a resemblance to the Me 163. It had a mid-set wing and was to be powered by a 3,750lb (1,700kg) st Walter rocket motor.

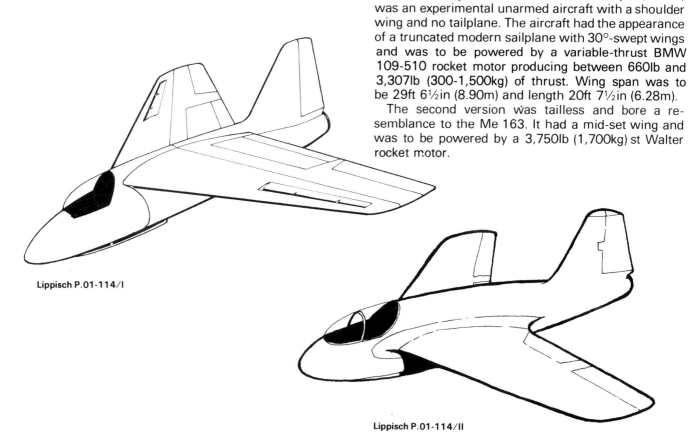

Lippisch P.01-114/I

Lippisch P.01-114/II

Lippisch P.01-115

The P.01-115 fighter was conceived on July 2, 1941, almost a year after the first P.01-114 design was produced. Again resembling the Me 163, it had mid-set wings with leading edges swept back at 35°. The fuselage had a short, rounded nose, with a single-seat cockpit set just behind it. Power was to be supplied by a single BMW 109-510 variable-thrust rocket and a 1,764lb (800kg) st BMW 003 turbojet fed by a dorsal air intake.

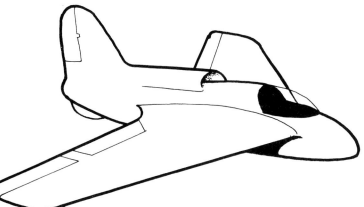

Lippisch P.01-115 data

Role	Tailless jet/rocket fighter
Ultimate status	Design
Powerplant	One BMW 003 turbojet (1,764lb, 800kg st) and one BMW 109-510 variable-thrust rocket motor (660-3,307lb, 300-1,500kg thrust)
Span	29ft 6½in (8.90m)
Length	24ft 7in (7.50m)
Armament	Two 30mm cannon

Lippisch P.01-116

Again, two versions of the one design seem to have been produced under the same designation. Curiously, the first P.01-116 preceded the P.01-111 by over six months, the design being dated April 13, 1939. An experimental tailless aircraft with extremely stubby, squared-off wings, the -116 had a flattened nose, indicating the possible presence of an air intake for a turbojet or ramjet. The cockpit canopy followed the contours of the fuselage and was located in the extreme nose. Wing span was 19ft 8in (6.0m) and length 17ft 9in (5.40m).

The second P.01-116 version, dated June 12, 1941, represented a clear step along the road towards the final Me 163 *Komet* shape. The aircraft was tailless and the mid-set wings were swept back at 35° at the leading edge. The single-seat cockpit was of conventional layout and located behind a short nose. The dual powerplant comprised a single 1,764lb (800kg) st BMW

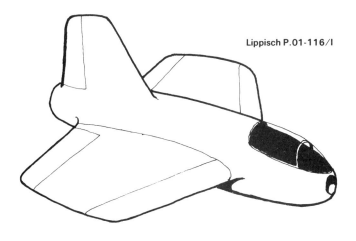

Lippisch P.01-116/I

turbojet positioned in the belly – the air intake was just below the cockpit and the jetpipe two-thirds of the way along the lower fuselage – and a 660-3,307lb (300-1,500kg) thrust BMW 109-510 variable-thrust rocket motor located in the rear fuselage above the turbojet exhaust. Armament was to be two 20mm cannon and two 13mm machine guns, and wing span was 29ft 6½in (8.90m) and length 23ft (7.0m).

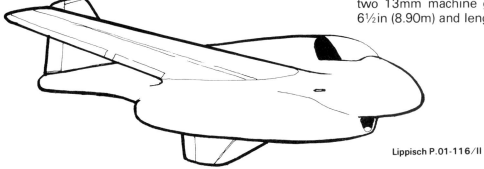

Lippisch P.01-116/II

Lippisch P.01-117

Another tailless design, the P.01-117 of July 22, 1941, was to have *Komet*-like wings with 30° sweep on the leading edges and a large central fin. The circular-section fuselage was cigar-shaped and fitted with a central landing skid, and the pilot lay prone behind the plexiglas nose cone.

Projected as a fighter, the -117 was to be armed with two 20mm cannon and two 13mm machine guns. Power was to be supplied by a variable-thrust BMW 109-510 rocket motor developing 660-3,307lb (300-1,500kg) of thrust. Span was 29ft 6½in (8.90m) and length 25ft (7.62m).

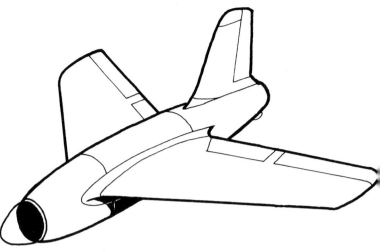

Lippisch P.01-118

By August 3, 1941, the long line of tailless fighter designs by Alexander Lippisch under the *Projekt X* programme was seen in the form of the P.01-118 to be nearing the shape of the Me 163 rocket interceptor which would ultimately enter operational service. Apart from the -118's extensions to the leading-edge wing roots, the planforms of the two aircraft are very similar, and it is only when they are viewed from the side that differences are clearly seen. The -118's fin was more rounded than that of the Me 163, and its cockpit canopy was simply a glazed area in the upper forward fuselage. Like its predecessors, the rocket-powered -118 was fitted with a central skid.

Lippisch P.01-118 data

Role	Tailless rocket fighter
Ultimate status	Design
Span	29ft 6½in (8.90m)
Length	23ft 6in (7.16m)
Armament	Two 30mm cannon

Lippisch P.01-119

Bearing a strong resemblance to the P.01-118, the P.01-119 was dated August 4, 1941, only a day after the P.01-118. A repositioned cockpit canopy seems to be the main difference between the two designs, the windscreen of the P.01-119 being reduced in length to result in a more conventional nose shape. Wing span and length were identical to those of the P.01-118, but the armament was increased to number four 20mm cannon.

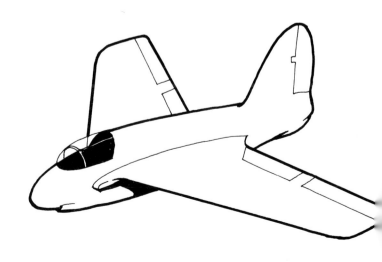

Lippisch Li 163S

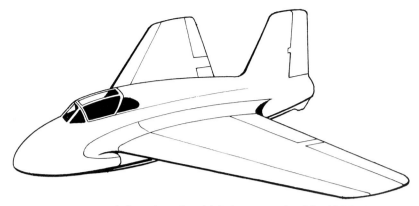

The first two Me 163A "V" (*Versuchs*, experimental) prototypes (see Messerschmitt Me 163, below) were completed in the spring of 1941. On October 2 of that year, during the early powered flight programme, a speed of over 623mph (1,003km/hr) was attained and the RLM ordered the development of the Me 163 as a fighter. While Messerschmitt was carrying out practical development, Lippisch had been busy working through his series of rocket-fighter designs towards the creation of the aircraft which became the Me 163B. It was in drawings dated September 14, 1941, that the unmistakable *Komet* shape finally emerged. Undoubtedly the direct predecessor of the Me 163B production rocket interceptor, this design was labelled the Lippisch Li 163. It was to have an armament of four 30mm cannon and a wing span of 30ft 2in (9.20m) and a length of 18ft 8½in (5.70m), very similar to the dimensions of the Me 163B.

Messerschmitt Me 163 Komet

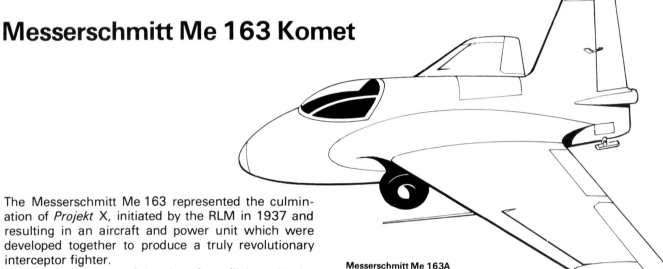

Messerschmitt Me 163A

The Messerschmitt Me 163 represented the culmination of *Projekt* X, initiated by the RLM in 1937 and resulting in an aircraft and power unit which were developed together to produce a truly revolutionary interceptor fighter.

Following a successful series of test flights with the DFS 194 (see page 43) at Peenemünde during 1940, Professor Lippisch's previously low-priority but highly secret *Projekt* X was speeded up. Three high-speed prototypes similar in layout to the DFS 194 were ordered. As the work had been moved to Messerschmitt from the DFS in early 1939, the three new aircraft were designated Messerschmitt Me 163.

The first two prototypes were completed in the spring of 1941. There followed a series of unpowered test flights during which a speed of 528mph (850km/hr) was attained in a dive. By August 1941 the Me 163 V1 had been fitted with a Walter HWK R.II-203 330-1,650lb (150-750kg) variable-thrust rocket motor. On only the fourth flight with this powerplant a level speed of 570mph (920km/hr) was attained, and on October 2, 1941, test pilot Heini Dittmar reached a speed of over 623mph (1,003km/hr) only a couple of minutes after casting off from the tow aircraft. The RLM immediately ordered the further development of the Me 163A interceptor. The first production prototypes of the *Komet* were completed in May 1942, and an initial batch of 70 Me 163Bs was ordered.

The first operational unit received a flight of Me 163Bs late in June 1944, and were first used operationally in July of that year. The first interception of American B-17 bombers followed on August 16, though none were shot down. Although 300 or so *Komets* were built, only around 11 successful attacks were recorded, and against this the type gained a reputation as a dangerous aircraft to fly and land. On landing the *Komet* tended to flip over on the belly skid, the wheels having been jettisoned on take-off, and remnants of the highly

Left: **Messerschmitt Me 163B**

Right: **Messerschmitt Me 163B-1a about to take off** (*via Pilot Press*)

Left: **Messerschmitt Me 163A V1** (*via Pilot Press*)

Below: **Messerschmitt Me 163A armed with R4M missiles** (*via Pilot Press*)

Messerschmitt Me 163 data

Role	Single-seat target-defence rocket interceptor
Ultimate status	Operational
Powerplant	One Walter HWK109-509A-2 rocket motor, 3,750lb (1,700kg) thrust (Me 163B) One Walter HWK 109-509C rocket motor (4,410lb, 2,000kg thrust) plus cruising chamber (660lb, 300kg thrust) (Me 163C)
Maximum speed	596mph at 9,840-29,530ft (960km/hr at 3,000-9,000m) (B)
Range	3min endurance (B), 12min (C)
Climb rate	16,400ft/min (5,000m/min)
Weight	4,200lb (1,905kg) empty, 9,500lb (4,310kg) loaded
Span	30ft 7in (9.30m) (B), 32ft 2in (9.80m) (C)
Length	18ft 8in (5.70m) (B), 23ft 1in (7.05m) (C)
Wing area	199.13ft² (18.5m²) (B), 219.58ft² (20.4m²) (C)
Armament	Two MK 108 30mm cannon or two MG 151/20 20mm cannon, augmented by five 50mm rocket shells fired vertically by light-sensitive cells from each wing root (SG 500 *Jagdfaust*)

volatile *T-Stoff* and *C-Stoff* fuel left in the tanks tended to explode in a horrifying manner when accidentally mixed.

The summer of 1944 saw the introduction of a new rocket motor, the HWK 509C, which supplanted the 3,750lb (1,700kg) thrust HWK 509A-2 fitted to the Me 163B. The new motor had a 660lb (300kg) thrust cruising chamber, and powered flight endurance was greatly increased when the new unit was incorporated into the redesigned Me 163C.

The Me 163C was slightly bigger than the B and more streamlined, with a bubble canopy and pressurised cockpit. Only a few Me 163Cs had been completed by the end of the war.

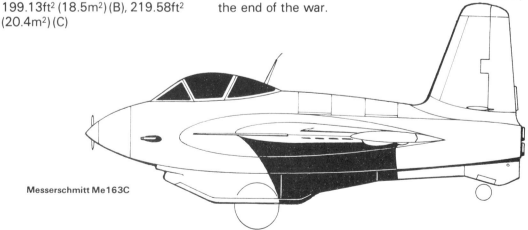

Messerschmitt Me163C

Messerschmitt Me 262

Messerschmitt Me 262A-1

Enjoying the distinction of being the world's first turbojet-powered fighter to reach operational status, the Me 262 had its origins in an RLM order placed in 1938 with Messerschmitt for an aircraft suitable for use with the new turbjojet engines being developed by BMW and Junkers. The project was designated P.1065 and a mock-up was inspected by RLM officials in December 1939, a contract for three prototypes following in March 1940.

The prototype airframes were ready well before the Junkers and BMW turbojets, which had for some reason received a lower priority, and so a 700hp Jumo 210G piston engine was installed in the nose of the Me 262 V1 for initial flight tests. Powered by this engine, the aircraft flew for the first time on April 18, 1941. The Me 262 was a very clean design, and featured slightly swept wings, the leading edges of which swept back at an angle of 18°. The low-set wings each carried an underslung turbojet located about six feet out from the fuselage sides. The tailplane was set high, nearly halfway up the fin. On production variants the cockpit was located centrally, under a bubble-type canopy. The nose was long and pointed, housing four MK 108 30mm cannon. The tailwheel undercarriage originally fitted to the first prototypes caused problems during take-off and was later replaced with a tricycle arrangement.

Eleven months after the first Jumo 210G-powered flight, two 1,000lb (454kg) st BMW 003 turbojets were fitted to the V1, but the first attempt to fly with the assistance of the Jumo 201G still installed in the nose failed after the turbojets suffered blade failure at take-

Messerschmitt Me 262A-1a of the Luftwaffe's first jet interceptor unit, the *Kommando Nowotny* (via Pilot Press)

Me 262B-1a/U1 night fighter fitted with *Liechtenstein* air-to-air radar

off. It was ultimately V2 which became the first Me 262 to fly on turbojet power only, taking to the air on July 18, 1942, fitted with a pair of 1,852lb (840kg) st Jumo 004s. Not longer after that another 45 Me 262s were ordered.

The first production models, designated Me 262A-1a, were named *Schwalbe* (Swallow) and entered operational service in May 1944. But then Hitler ordered that all the A-1a interceptors should be converted to attack bombers. Resulting in the mounting of a pair of 550lb bombs beneath the nose, this immediately had a detrimental effect on their performance. At the same time the type was renamed *Sturmvogel* (Stormbird) and designated Me 262A-2a.

There followed a multitude of Me 262 variants built in small numbers, including the two-seat Me 262B for training and a night-fighter modification of the B as the Me 262B-la/U1. Several Me 262C interceptors (*Abfangjäger*) boosted by tail-mounted bi-fuel rocket motors were proposed, and the prototype Me 262C-1a first flew on February 27, 1945. The Me 262C-2b was to have BMW 003R engines comprising a 1,764lb (800kg) st BMW 003A turbojet and a 2,650lb (1,200kg) thrust BMW 109-718 bi-fuel rocket motor in each wing nacelle. But after only one flight with this arrangement attention was turned to the Me 262C-3 with underfuselage jettisonable rocket motors; the war ended before this variant could be tested.

Proposals for three unarmed reconnaissance variants were also produced. The first, the *Aufklärer* I, had a transparency in the cockpit floor and was powered by two Jumo 004C turbojets. Maximum range was 956 miles at 29,530ft (1,540km at 9,000m) on 670 Imp gal (3,045lit) of fuel. Two cameras, an Rb 75/30 and an Rb 20/30, were to be mounted in the nose. The *Aufklärer* la was to have two Rb 75/30 cameras and be armed with two MK 108 30mm cannon. The cockpit was located in the nose and the cameras in the fuselage behind the rear fuel tank. The third *Aufklärer* type, the II, was a modified Me 262A-2. Fuselage depth was increased to accommodate a 319 Imp gal (1,450lit) internal tank, two auxiliary take-off wheels were fitted, and fin area was to be increased. Tankage was increased to 1,200 Imp gal (5,455lit), giving a range at 100 per cent thrust of 1,467 miles (2,360km).

Three fast bomber proposals – *Schnellbomber* I, la and II – outwardly resembled their reconnaissance counterparts, but had the latter's extra fuel weight allowance given over to bomb load.

Further proposed variants included a Me 262 powered by Jumo 004Cs, and two versions with 35° and 45° sweepback respectively and powered by HeS 011s faired into the wing roots. It was also proposed to fit two Sänger ramjets to an Me 262 to supplement the Jumo 004Bs.

Of the 1,433 Me 262s built, only a little over 200 actually attained operational status. Though the Me 262 was a highly promising type, its effectiveness was reduced by Hitler's insistence that it first be used in the bombing role, as well as the chronic shortage of fuel, chaotic supply situation, and underdeveloped combat tactics.

Above: Captured Me 262s photographed at Lechfeld, Germany, in June 1945 (*USAF*)

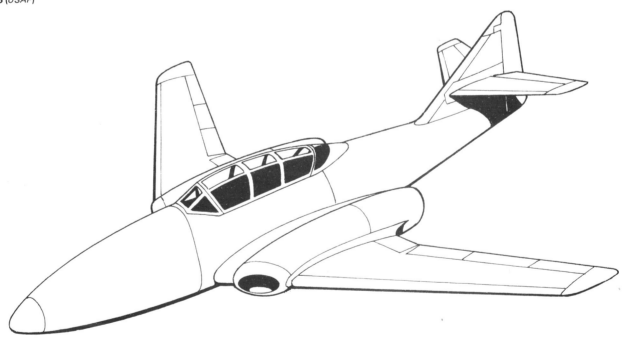

Swept-wing Me 262

Messerschmitt Me 262A-1a data

Role	Single-seat jet fighter
Ultimate status	Operational
Powerplant	Two Junkers Jumo 004B-1 turbojets, 1,984lb (900kg) st each
Maximum speed	538mph at 29,530ft (866km/hr at 9,000m)
Range	652 miles at 29,530ft (1,050km at 9,000m)
Weight	9,744lb (4,420kg) empty, 14,101lb (6,395kg) loaded
Span	40ft 11½in (12.48m)
Length	34ft 9½in (10.60m)
Wing area	234ft² (21.7m²)
Armament	Four MK 108 30mm cannon and 24 R4M 55mm rockets (optional)

Me 262 proposals

	Powerplant	Loaded weight lb (kg)	Speed at height mph, feet (km/hr, m)	Range miles (km)	Armament
Abfangjäger I	2×Jumo 004C 1×HWK R.II-211	17,604 (7,985)	545 at 38,400 (877 at 11,700)	463 (745)	6×MK 108 30mm cannon
Abfangjäger II	2×BMW 003R or 2×BMW 004B	15,600 (7,075)	460 at 38,400 (740 at 11,700)	525 (845)	5×MK 108 30mm cannon
Abfangjäger III	2×HWK R.II-211	15,300 (6,940)	—	193 (310)	6×MK 108 30mm cannon
Aufklärer I	2×Jumo 004C	14,506 (6,580)	590 at 29,530 (950 at 9,000)	957 (1,540)	—
Aufklärer Ia	2×Jumo 004C	14,506 (6,580)	590 at 29,530 (950 at 9,000)	957 (1,540)	2×MK 108 30mm cannon
Aufklärer II	—	20,000 (9,070)	547 at 19,690 (880 at 6,000)	1,467 (2,360)	—
Schnellbomber I	2×Jumo 004C	18,300 (8,300)	488 at 19,690 (785 at 6,000)	970 (1,560)	1,984lb (900kg) of bombs
Schnellbomber Ia	2×Jumo 004C	—	488 at 19,690 (785 at 6,000)	—	1,984lb (900kg) of bombs, 2×MK 108 30mm cannon
Schnellbomber II	—	20,000 (9,070)	547 at 19,690 (880 at 6,000)	1,120 (1,800)	1,984lb (900kg) of bombs, 2×MK 108 30mm cannon
Me 262 (004C)	2×Jumo 004C	14,176 (6,430)	590 at 26,250 (950 at 8,000)	—	—
Me 262 (swept wing)	2×HeS 011	—	590 (950)	—	—
Me 262 (ramjets)	2×Jumo 004B, 2×ramjets	—	620 (1,000) at sea level	—	—

Messerschmitt Me 263A

Known at one point as the Ju 248, the Me 263 was a Junkers development of the Me 163B and C rocket-propelled interceptors. It had a much slimmer fuselage of semi-monocoque construction and built in three main sections. A pressurised cockpit with armour protection for the pilot was riveted to the main fuselage section, while the rear portion was detachable to facilitate easy inspection and maintenance of the HWK 109-509C rocket motor.

As Me 163 development proceeded it became clear that the landing and stability problems could not be easily eradicated from the type in its existing form. Consequently, early in 1944 design work started on a radically improved version known initially as the Me 163D. At an early stage the work was handed over to Junkers, and the result, finally designated Me 263, was an interceptor which went some way towards eliminating the Me 163's landing and endurance problems. There was provision for more fuel and a retractable tricycle undercarriage, though the wings and fin were similar to those of the Me 163B.

During the initial unpowered trials the undercarriage was kept in the fixed position, and while the Me 263 proved to be much safer than the *Komet* at lower speeds, the centre of pressure moved back rapidly at speeds above Mach 0.8 and it could not be flown safely at speeds above 590mph (950mk/hr).

The rocket motor had an auxiliary combustion chamber mounted below the main unit to give a more econmical cruise. The *C-Stoff* and *T-Stoff* fuel tanks were emptied in flight in a prearranged order to minimise centre-of-gravity variations.

Preparations for large-scale production of the Me 263A1 were under way when the war ended.

Messerschmitt Me 263 data

Role	Single-seat target-defence rocket interceptor
Ultimate status	Flight test
Powerplant	Walter HWK 109-509C rocket motor, 4,410lb (2,000kg) thrust
Maximum speed	621mph at 9,840-36,090ft (1,000km/hr at 3,000-11,000m)
Range	90 miles at 20,670ft (145km at 6,300m)
Endurance	13.2min
Weight	4,640lb (2,105kg) empty, 11,695lb (5,305kg) loaded
Span	31ft 2½in (9.50m)
Length	25ft 11½in (7.95m)
Wing area	191.6ft² (17.8m²)
Armament	Two MK 108 30mm cannon

Messerschmitt Me 263 V1 *(IWM)*

Messerschmitt Me 264

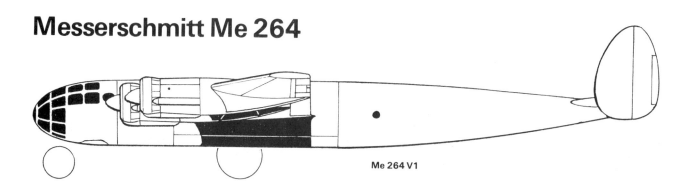

Me 264 V1

Originally designed for a range of up to 9,320 miles (15,000km) on the power of four BMW 801E piston engines, this bomber and reconnaissance aircraft was also proposed with jet or mixed powerplants.

In 1941 the RLM studied the possibility of an *Amerika-Bomber* to be used against the United States in the event of that country's entering the war, and ordered three prototypes from Messerschmitt. The first, the Me 264 V1, was flown for the first time late in 1942 on the power of four Jumo 211J piston engines. The extremely clean design was characterised by a high shoulder wing, tricycle undercarriage and high-set tailplane with twin oval fins and rudders. The lines of the circular-section fuselage were unbroken by a cockpit canopy, the crew being accommodated behind a glazed nose section.

The project continued at a leisurely pace until early 1944, during which time alternative versions featuring various powerplants were proposed. These included a variant with the basic four BMW 801s plus a Jumo 004 turbojet housed in the rear of each of the outer engine nacelles. Estimated maximum speed was increased to 407mph at 21,990ft (655km/hr at 6,700m). Jet-only proposals included one with four turbojets and another designed to tow an Me 328 pulsejet-powered fighter. Other variants were to have just two BMW 018 turbojets or two BMW 028 turboprops. A lengthened version with four BMW 801TGs and two BMW 018s was also considered, as was a steam turbine-powered version. This was commissioned in August 1944, the powerplant to be designed and developed by Professor Losel of Osermaschinen GmbH. Intended to develop 6,000hp at 6,000rpm, the turbine was to drive one of two forms of propeller: either 17ft 6in (5.35m) in diameter and turning at 400-500rpm, or 6ft 6½in (2.0m) in diameter and turning at 6,000rpm. Many parts had been made by the time of the German collapse, but the Me 264 airframe which had been placed at the Osermaschinen's disposal had been destroyed in an air raid.

Messerschmitt Me 264 data

Role	Mixed-power six-seat long-range reconnaissance bomber
Ultimate status	Construction
Powerplant	Four BMW 801 piston engines (standard), plus two Jumo 004 turbojets (1,984lb, 900kg st each); or two BMW 018 turbojets (7,496lb, 3,400kg st each) or two BMW 028 turboprops (6,570eshp each); or four BMW 801s and two BMW 018s
Maximum speed	373mph at 34,440ft (600km/hr at 10,500m) (4×BMW 801TG) 407mph at 21,990 (655km/hr at 6,700m) (4×BMW 801 + 2× Jumo 004) 484mph at 22,960ft (780km/hr at 7,000m) (2×BMW 018) 500mph at 21,990ft (805km/hr at 6,700m) (2×BMW 028) 494mph at 23,950ft (795km/hr at 7,300m) (4×BMW 801 + 2× BMW 018)
Range	9,315 miles (14,990km), 45hr endurance (4 × BMW 801)
Weight	51,520lb (23,365kg) empty, 100,416lb (45,550kg) loaded
Span	141ft (43.0m)
Length	68ft 7in (20.90m)
Wing area	1,377.8ft² (128.0m²)
Armament	Two MG 151/20 20mm cannon and four MG 131 13mm machine guns in five fuselage points, and 4,410lb (2,000kg) of bombs

Messerschmitt Me 264 V1 (*IWM*)

Messerschmitt Me 328

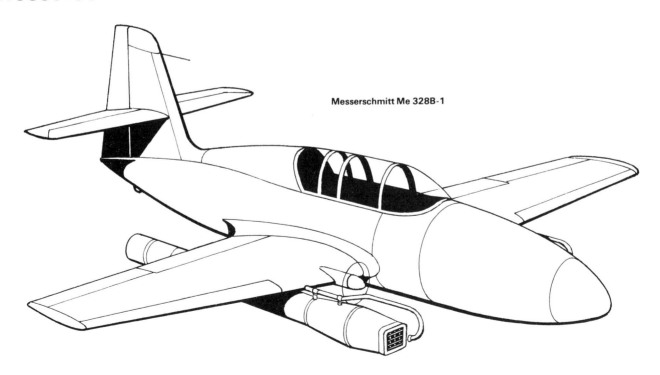

Messerschmitt Me 328B-1

Design work on the experimental Me 328 started late in 1942. It was intended as a cheap, high-speed, low-level bomber or emergency day fighter and was to become one of the few piloted aircraft to fly under the power of pulsejets alone. By March 1943 the pace of development had been intensified and a prototype had undergone initial tests mounted on the back of a Do 217

Made mostly of wood, the Me 328 had low/mid-set straight wings and a tailplane mounted about halfway up the single fin. The pilot was accommodated in a single centrally positioned cockpit behind the long, pointed nose.

Although initial tests revealed that the Me 328 had unsatisfactory aerodynamics, development proceeded nevertheless. Early flight trials were marked by several failures in the rear structure. The culprit was cost-cutting, for, in order to speed production and minimise the number of skilled workers needed, the Me 328 was to be powered by a pair of Argus pulsejets, even though the acoustic effects of this powerplant were known to be very damaging to the airframe.

Some ten prototypes of the Me 328 were produced, with varying powerplant arrangements. The Me 328A-1/2 escort fighter had two wing-mounted pulsejets, and the other escort design, the Me 328A-2, had no fewer than four fuselage-mounted pulsejets. The escort fighters were to be towed into the air and released to fight with their full endurance remaining. But such was the rapid dropping-off in efficiency of the Argus pulsejets at normal aircraft operational heights that this scheme was abandoned in favour of the Me 328B fast bomber, which could be carried to the target in a *Mistel* (composite) arrangement. The Me 328B had assisted take-off, using either a catapult or short-duration rockets, landings being made on retractable fuselage skids.

The problems of acoustic damage from the pulsejets were never fully solved, and the project had to be temporarily rescued by fitting a Jumo 004B turbojet, resulting in the Me 328C. This put the price up unacceptably, however, and eventually an unpowered version of the Me 328B was ordered into production as a piloted glide bomb. But even this was dropped in favour of the Fi 103 *Reichenberg*, and ultimately no production Me 328s were built.

Messerschmitt Me 328 data

Role	Pulsjet-powered escort fighter (Me 328A), low-level bomber (Me 328B)
Ultimate status	Flight test
Powerplant	Two Argus As 014 pulsejets (661lb, 300kg thrust each) (A-1/2) Four Argus As 014 pulsejets (661lb, 300kg thrust each) (A-2) Two Argus As 014 pulsejets (661lb, 300kg thrust each) (B) One Junkers Jumo 004B turbojet (1,984lb, 900kg st) (C)
Maximum speed	472mph (760km/hr) at sea level (A-1/2), 572mph (920km/hr) at sea level (A-2), 423mph (680km/hr) at sea level with external bomb load (B)
Range	478 miles (770km) (A-1/2), 870 miles (1,400km) (A-2), 392 miles (630km) (B)
Weight	4,840lb (2,185kg) (A-1/2), 8,355lb (3,790kg) (A-2), 5,941lb (2,695kg) (B)
Span	21ft (6.40m) (A-1/2 and B), 27ft 10½in (8.50m) (A-2)
Length	22ft 5in (6.83m) (A-1/2 and B), 28ft 3in (8.60m) (A-2)
Wing area	80.7ft² (7.5m²) (A-1), 129.1ft² (12.0m²) (A-2), 101.2ft² (9.4m²) (B)
Armament	Two MG 151/20 20mm cannon (A-1/2 and 2) plus 1,102lb (500kg) of bombs (B)

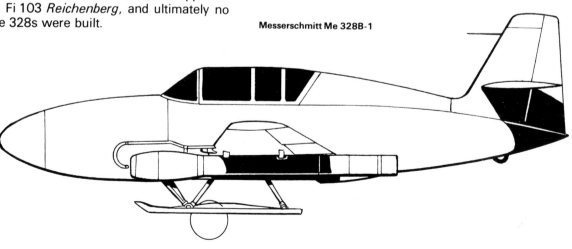

Messerschmitt Me 328B-1

Messerschmitt Me 362

An intriguing but poorly documented project, the Me 362 was designed as a jet airliner. It was probably a private venture with an eye to a future in which quantity production of military aircraft had dropped off after the war. All that is known is that it was to be powered by three turbojets.

Messerschmitt P.1065

In 1938 the RLM placed an order with Messerschmitt for the design of an airframe capable of receiving the first generation of turbojet engines. Initially the project was designated P.1065, but when an order for three airframes was placed by the RLM in 1940 the project was redesignated Me 262 (see Me 262, page 116).

Messerschmitt P.1079

Turbojets were very much an untried form of power during the war, and the best form of installation for them was still largely a matter of conjecture. Experimental data were lacking, and so the single-seat P.1079 was designed for the study of various jet installations. Provision was to be made for two turbojets.

Messerschmitt P.1092

Three versions of the P.1092 were designed, all of them fighters powered by single Junkers Jumo 004C turbojets. The first, the P.1092A, was to have sharply swept-back wings and tailplane with an aspect ratio of 4.75 to 1, and a wing span of 25ft 6in (7.77m). Two other versions, the P.1092B and B2, were similar to the A but had a larger wing span of 32ft 9in (9.98m) and an aspect ratio of 7 to 1. The latter figure suggests that the wings may have been straight or at least not as sharply swept as in the A; moreover, contemporary descriptions do not emphasise the wing shape in the same manner as they do that of the P.1092A. All three types had the same length but performance, weights and other details differed in varying degrees.

Messerschmitt P.1092 data

Role	Single-seat jet fighter
Ultimate status	Design
Powerplant	One Jumo 004C turbojet, 2,205lb (1,000kg) st
Maximum speed	575mph (925km/hr) (A), 572mph (920km/hr) (B), 578mph (930km/hr) (B2), all at 26,250ft (8,000m) 541mph (870km/hr) (A), 528mph (850km/hr) (B and B2), all at sea level
Weight	4,453lb (2,020kg) (A), 4,612lb (2,092kg) (B and B2), all empty 5,776lb (2,620kg) (A), 5,930lb (2,690kg) (B), 5,148lb (2,335kg) (B2) loaded
Span	25ft 7½in (7.80m) (A), 32ft 10in (10.0m) (B and B2)
Length	27ft 3in (8.30m)
Wing area	136.7ft^2 (12.7m^2) (A), 155.4ft^2 (14.4m^2) (B and B2)
Armament	Two MK 103 30mm cannon and two MG 151/15 15mm machine guns (A), two MK 103 30mm cannon and two MG 151/15 15mm machine guns (B), one MK 103 30mm cannon and two MG 151/15 15mm machine guns (B2)

Messerschmitt P.1099

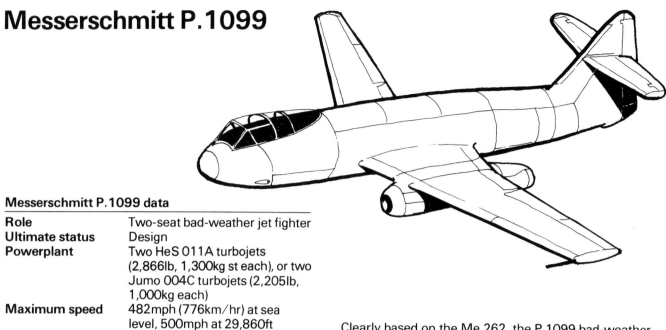

Messerschmitt P.1099 data

Role	Two-seat bad-weather jet fighter
Ultimate status	Design
Powerplant	Two HeS 011A turbojets (2,866lb, 1,300kg st each), or two Jumo 004C turbojets (2,205lb, 1,000kg each)
Maximum speed	482mph (776km/hr) at sea level, 500mph at 29,860ft (805km/hr at 9,100m) (with Jumo 004Cs)
Range	814 miles at 19,690ft (1,340km at 6,000m)
Ceiling	32,320ft (9,810m)
Weight	19,004lb (8,620kg) (maximum), 13,404lb (6,080kg) (landing)
Span	41ft 4½in (12.60m)
Length	39ft 4in (12.0m)

Clearly based on the Me 262, the P.1099 bad-weather fighter project was to have a deeper fuselage than the Me 262 to accommodate additional fuel tanks and/or armament. The two-seat cockpit was located well forward in the nose and the aircraft was to be powered by two Jumo 004C turbojets or two Heinkel HeS 011s. The fixed armament consisted of either three or four 30mm cannon, or two 30mm cannon, or one 30mm cannon and one 55mm cannon with 40 rounds for the larger weapon.

Messerschmitt P.1100

Based on the Me 262A-2, the P.1100 had a deeper fuselage to permit bombs to be carried internally and was probably similar in appearance to the Me 262 *Schnellbomber* II. Intended for use as a fast bomber, the P.1100 was unarmed and the crew of two were seated side by side, with provision for an extra rear-facing seat for an observer. The powerplant, like that of the P.1099, was to be two Jumo 004C turbojets, though the installation of two HeS 011s was also foreseen.

Messerschmitt P.1100 data

Role	Two or three-seat fast jet bomber
Ultimate status	Design
Powerplant	Two Jumo 004C turbojets (2,205lb, 1,000kg st each) or two HeS 011A turbojets (2,866lb, 1,300kg st each)
Maximum speed	486mph (780km/hr) at sea level, 515mph (830km/hr) at 19,690ft (6,000m)
Service ceiling	30,840ft (9,400m)
Range	825 miles at 22,965ft (1,330km at 7,000m)
Weight	20,084lb (9,110kg) (loaded), 12,103lb (5,490kg) landing
Armament	5,512lb (2,500kg) of bombs

Messerschmitt P.1101

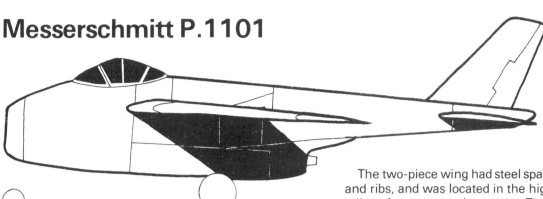

Early in 1944 the Messerschmitt advanced design office at Oberammergau began designing and developing a single-jet fighter with approximately the same endurance and range as the Me 262. Of the several projects considered, the P.1101 was produced at the personal request of Willy Messerschmitt for research into high-speed flight and experiments with swept wings. After a quarter-scale model had been wind-tunnel-tested at Berlin-Adlershof, an experimental prototype was started in July 1944. It was approaching completion as the war drew to a close, and incorporated a number of advanced features, including a facility for altering the wing sweep on the ground through angles from 35° to 45°. The arrival of Allied troops forestalled the first flight, and an attempt to destroy the prototype was made before it was finally captured.

The prototype was fitted with a Jumo 004B engine but it was expected to use the HeS 011 unit on production versions in the event of its winning the Emergency Fighter competition. For this submission the wing sweep angle was fixed at 40°. It was not in fact selected by the OKL but work continued on the prototype, which was later used by the Americans as a design basis for their Bell X-5 variable-sweep research aircraft.

The two-piece wing had steel spars with wooden skin and ribs, and was located in the high-mid position; the tail surfaces were also swept. The pressurised cabin was located well forward in the upper part of the fuselage, and behind it were the fuel tank and then space for the retracted main undercarriage legs. Below the cockpit was the air intake and duct for the turbojet, which was mounted in mid-fuselage. The armament of two or four 30mm cannon was disposed on either side of the cockpit.

Messerschmitt P.1101 data

Role	Single-seat jet research aircraft and proposed Emergency Fighter
Ultimate status	Construction
Powerplant	One Junkers Jumo 109-004B turbojet (1,984lb, 990kg st) (prototype), one HeS 011A turbojet (2,866lb, 1,300kg st) (Emergency Fighter)
Maximum speed	549mph (885km/hr) at sea level, 609mph (980km/hr) at 22,965ft (7,000m)
Range	932 miles (1,500km)
Ceiling	45,930ft (14,000m)
Weight	6,065lb (2,750kg) empty, 9,526lb (4,320kg) loaded
Span	27ft 1in (8.25m)
Length	30ft 4in (9.24m)
Wing area	170.6ft² (15.85m²)
Armament	Two to four MK 108 cannon

Messerschmitt P.1101 mock-up photographed at Oberammergau, Germany, in 1945 (*via Pilot Press*)

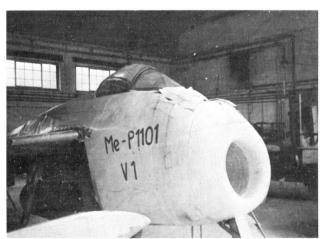

Messerschmitt P.1102

This project was designed in the summer of 1944 as an unarmed jet bomber powered by three 1,764lb (800kg) st BMW 003 turbojets. The sweep angle of the mid-set wings could be adjusted through an arc of 30°, from 20° to 50°, and the tail surfaces were also sharply swept. The engines were mounted one on either side of the lower nose in external pods, and the third exhausting from the tail; it is not clear where the air intake for the third unit was located.

Messerschmitt P.1104

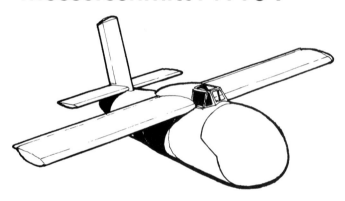

Designed specifically as a target-defence interceptor, the diminutive P.1104 did not progress beyond the design stage. Its layout was very simple, with straight shoulder wings only 70ft² in area and squared-off tail surfaces. The cigar-shaped fuselage housed a single Walter HWK 109-509A rocket motor exhausting at the tail, and a centrally located single-seat cockpit. It was intended to enter operational service by August 1945, taking off from a 160yd (150m) horizontal surface and landing on a central retractable skid. Armament was to consist of a single 30mm cannon.

Messerschmitt P.1104 data

Role	Single-seat rocket target-defence interceptor
Ultimate status	Design
Powerplant	One Walter HWK 109-509A-2 rocket motor, 3,750lb (1,700kg) thrust
Maximum speed	497mph (800km/hr)
Range	53 miles (85km) after climbing to 19,690ft (6,000m) at 39,370ft/min (12,000m/min)
Weight	5,655lb (2,565kg) loaded
Wing area	70ft² (6.5m²)
Armament	One MK 108 30mm cannon, 100 rounds

Messerschmitt P.1106

Intended as an improvement on the P.1101, the very similar P.1106 was not developed beyond the design stage. It differed from the P.1101 mainly in having the cockpit repositioned just in front of the tail, the vacated volume being taken up by two 30mm cannon and the tricycle undercarriage. This arrangement resulted in a poor view for the pilot and very little improvement in performance, and consequently the P.1106 was abandoned.

The first design, with a T-tail, incorporated the pilot's cockpit into the fin; in the second design, with a butterfly tail, the pilot's cockpit had a completely separate canopy.

Messerschmitt P.1106 data

Role	Single-seat jet fighter
Ultimate status	Design
Powerplant	One HeS 011A turbojet, 2,866lb (1,300kg) st
Maximum speed	677mph at 19,690ft (1,091km/hr at 6,000m) (T-tail), 618mph at 22,965ft (995km/hr at 7,000m) (butterfly tail)
Range	994 miles (1,600km) (T-tail)
Weight	8,362lb (3,793kg) loaded (T-tail), 8,818lb (4,000kg) (butterfly tail)
Span	22ft 2in (6.74m) (T-tail), 21ft 10in (6.65m) (butterfly tail)
Length	26ft 3in (8.00m) (T-tail), 29ft 10½in (9.10m) (butterfly tail)
Wing area	139.9ft² (13.0m²) (T-tail), 135.2ft² (12.56m²) (butterfly tail)
Armament	Two MK 108 30mm cannon

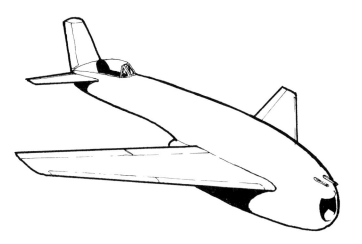

Messerschmitt P.1107

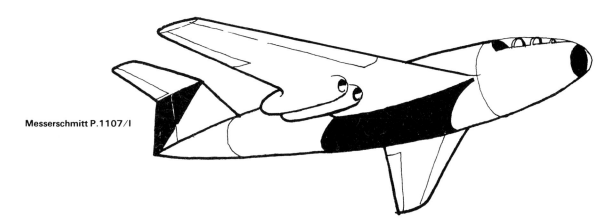

Messerschmitt P.1107/I

Projected early in 1945 to compete with the Horten XVIII and Junkers EF.130, the P.1107 long-range jet bomber was designed in two versions. Both had cigar-shaped fuselages unbroken by cockpit canopies, all transparent surfaces being faired into the fuselage form. The mid-set wings were swept back and a tricycle undercarriage retracted rearwards into the fuselage.

Both powered by four HeS 011 turbojets, the two versions differed in tail unit arrangement and engine location. The engines of the P.1107/I were slung beneath the wing and mounted in twin nacelles with a separate circular intake for each unit; the jetpipes extended well beyond the wing trailing edge. The tailplane was mounted high on top of the fin, and both were slightly swept.

On the P.1107/II the turbojets were mounted in the wing, close into the fuselage sides, with a single elongated air intake in the wing leading edge feeding each pair of engines. The butterfly tailplane was moderately swept.

Messerschmitt P.1107 data

Role	Four-seat long-range jet bomber
Ultimate status	Design
Powerplant	Four HeS 011A turbojets, 2,866lb (1,300kg) st each
Maximum speed	580mph at 22,970ft (933km/hr at 7,000m) (P.1107/II)
Weight	68,000lb (30,854kg) loaded, 40,000lb (18,145kg) empty (P.1107/I)
	67,505lb (30,620kg) loaded, 36,990lb (16,780kg) empty (P.1107/II)
Span	57ft 1½in (17.40m) (both versions)
Length	60ft (18.30m) (P.1107/I), 59ft (18.0m) (P.1107/II)
Wing area	655ft² (60.85m²) (both versions)
Armament	8,820lb (4,000kg) of bombs

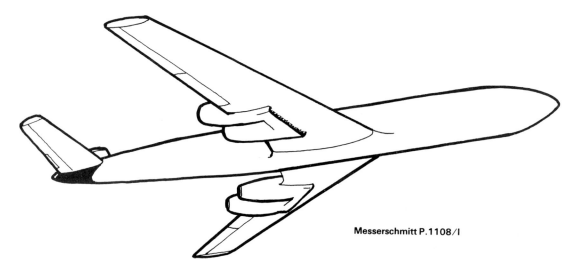

Messerschmitt P.1107/II

Messerschmitt P.1108

Messerschmitt P.1108/I

Two designs for a medium jet bomber bore the P.1108 designation. They were intended to meet a "1,000 × 1,000 × 1,000" specification, being designed to carry a bomb load of 1,000kg over a range of 1,000km at 1,000km/hr.

The first version had low, 30°-swept wings and a butterfly tail, and was powered by four Heinkel HeS 011 turbojets mounted in a similar fashion to those of the P.1107/I. In all, this version looked very much like a combination of the P.1107/I and II.

The tailless P.1108/II had delta wings and a single central fin. It was also to be powered by four HeS 011 turbojets mounted in the wings, with elongated leading-edge intakes feeding each unit.

Messerschmitt P.1108 data

Role	Medium jet bomber
Ultimate status	Design
Powerplant	Four HeS 011A turbojets, 2,866lb (1,300kg) st each
Maximum speed	621mph (1,000km/hr)
Range	1,498 miles (2,410km)
Span	69ft 6in (21.18m) (P.1108/II)
Length	59ft 8½in (18.20m) (P.1108/II)
Armament	4,410lb (2,000kg) of bombs

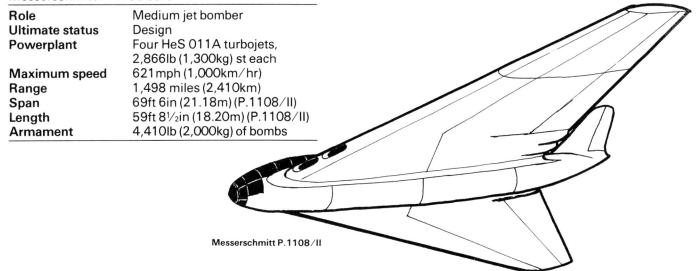

Messerschmitt P.1108/II

Messerschmitt P.1110

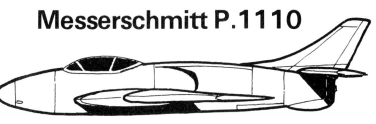

Messerschmitt P.1110/I

This project was an effort to establish the optimum airframe location for jet engines and other equipment in the quest for maximum aerodynamic efficiency. Fuselage cross-sectional area was one of the parameters looked at most closely as the Messerschmitt design team anticipated by about ten years the work of the American Dr Richard Whitcomb on the transonic area rule.

Two versions, with similar nose and wings, were designed under the P.1110 designation. The first, with 40°-swept wings similar to those of the P.1101, was submitted for the Emergency Fighter competition but was not accepted. A conventional swept tail unit was fitted and the air intakes for the single turbojet engine were positioned on either side of the fuselage, just aft of the cockpit.

The P.1110/II differed from the wing trailing edge back, the air intake being an annular slot running round the whole circumference of the fuselage, while the tail unit was changed to a butterfly arrangement. Behind the pressurised cabin was a self-sealing fuel tank, with the HeS 011 turbojet mounted in the tail. The armament of three MK 108 cannon was housed in the nose, with room for two more if necessary.

Messerschmitt P.1110 data

Role	Single-seat jet research aircraft and Emergency Fighter
Ultimate status	Design
Powerplant	One HeS 011A turbojet, 2,866lb (1,300kg) st
Maximum speed	617mph (993km/hr) (I), 625mph (1,006km/hr) (II), both at 22,970ft (7,000m)
Range	932 miles (1,500km)
Ceiling	42,650ft (13,000m)
Weight	9,440lb (4,282kg) (I), 8,818lb (4,000kg) (II) loaded
Span	27ft 1in (8.25m) (I), 21ft 10½in (6.65m) (II)
Length	33ft 11½in (10.35m) (I), 27ft 10½in (8.50m) (II)
Wing area	170.6ft² (15.85m²) (I), 140.5ft² (13.05m²) (II)
Armament	Three to five MK 108 30mm cannon

Messerschmitt P.1110/II

Messerschmitt P.1111

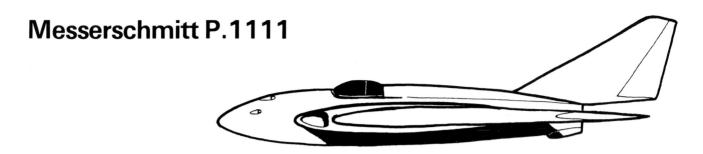

Designed for the Emergency Fighter competition of 1944, the P.1111 was a tailless project with swept delta wings. The deep-chord wing centre sections had a leading-edge sweep of 45° and slight sweepback on the trailing edges, as did the fin. A single fuselage-mounted HeS 011 turbojet was fed by an air intake in each leading-edge wing root, and the whole fuel supply of 330 Imp gal (1,500 lit) was contained in the wings. The project was dropped in favour of the P.1112.

Messerschmitt P.1111 data

Role	Single-seat tailless jet fighter
Ultimate status	Design
Powerplant	One HeS 011A turbojet, 2,866lb (1,300kg) st
Maximum speed	559mph (900km/hr) at sea level, 618mph at 22,970ft (995km/hr at 7,000m)
Range	932 miles (1,500m)
Service ceiling	45,930ft (14,000m)
Weight	6,064lb (2,750kg) empty, 9,440lb (4,282kg) loaded
Span	30ft 0½in (9.15m)
Length	21ft 0½in (6.40m)
Wing area	301.4ft² (28.0m²)
Armament	Four MK 108 30mm cannon

Messerschmitt P.1112

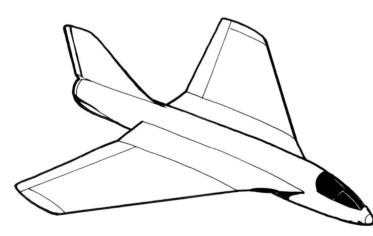

The design faults which became apparent in the P.1111 were rectified with the P.1112. Again tailless, it had near-delta wings, the area of which was reduced by about 20 per cent by comparison with the P.1111 because it was thought that the wing loading could profitably be increased. This project's lines were generally cleaner than those of the P.1111, with the pilot seated in the extreme nose behind a finely contoured cockpit cover. The windscreen was to be of 100mm-thick laminated glass. The power unit was the same as that of the P.1111, though the wing-root air intakes were elongated and set slightly out from the fuselage sides.

Messerschmitt P.1112 data

Role	Single-seat jet fighter
Ultimate status	Design
Powerplant	One HeS 011A turbojet, 2,866lb (1,300kg) st
Maximum speed	684mph (1,100km/hr)
Weight	10,295lb (4,670kg) loaded
Span	30ft 0½in (9.15m)
Length	31ft 6½in (9.60m)
Wing area	258.3ft² (24.0m²)
Armament	Two MK 108 30mm cannon

Messerschmitt P.1113

Very little is known about the P.1113, which was referred to simply as a tailless jet fighter powered by a single turbojet. No further information is available, but it seems reasonable to assume that it was a further development of the P.1111 and P.1112.

Messerschmitt P.1116

This project was similar in outline to the P.1101, though the cockpit was set well back in the fuselage, behind the fuel tank. Designed to the Emergency Fighter specification of September 1944, it was to have 40°-swept wings and a butterfly tail, and was expected to attain Mach 0.882. The single HeS 011 turbojet was to be mounted centrally in the lower fuselage, in a position similar to that of the P.1101, and fed by a nose intake. This arrangement did not improve performance, however, and the pilot's view was worse.

Messerschmitt P.1116 data

Role	Single-seat jet fighter
Ultimate status	Design
Powerplant	One HeS 011A turbojet, 2,866lb (1,300kg) st
Maximum speed	615mph at 22,970ft (990km/hr at 7,000m)
Weight	8,818lb (4,000kg) at take-off
Span	21ft 9¾in (6.65m)
Length	29ft 10½in (9.10m)
Wing area	135ft^2 (12.56m^2)

Opel-Hatry Rak-1

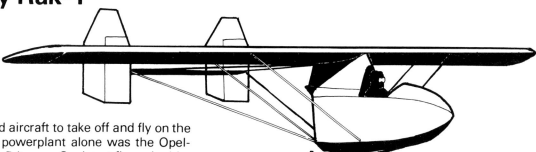

The very first manned aircraft to take off and fly on the power of a reaction powerplant alone was the Opel-Hatry Rak-1, built by Fritz von Opel and flown by him in September 1929. Little more than a high-winged glider with 16 small solid-fuel rocket motors mounted in the rear of its short fuselage, the Rak-1 took off from a 70ft-long raised wooden track.

Development of the aircraft was initiated by Max Valier, a key member of the VfR (Society for Space Navigation), who wanted a demonstration aircraft with which to raise funds for the development of a liquid-fuel rocket motor. At that time, 1928, the only rocket motors worth using were little powder units built by a pyrotechnicist named Alexander Sander. Fritz von Opel was intrigued by the idea of rocket power and teamed up with Valier and Sander to produce a rocket car as well as an aircraft. Opel engaged two designers, Lippisch and Hatry, to build a suitable aircraft. Hatry's design was the more promising, although it flew more than a year after Lippisch's catapult-launched canard aircraft.

After a number of false starts the first flight took place on September 30, 1929. It lasted ten minutes, the rockets being fired in relays. Towards the end of a planned series of flights with more powerful rockets fitted, the wings caught fire while the aircraft was still airborne and forced a landing on the skid undercarriage at the dangerously high speed of 70mph (112km/hr). Opel, who was flying the Rak-1 at the time, escaped the consequent crash with few injuries. But the aeroplane was badly damaged and Opel considered the project to be at an end, having extracted from it all the publicity he needed.

Opel-Sander Rak-1 data

Role	Rocket-propelled experimental glider
Ultimate status	Flight test
Powerplant	16 solid-fuel rocket motors, 44lb (20kg) thrust each
Maximum speed	95mph (153km/hr)
Endurance	10min
Weight	595lb (270kg)

Sänger-Bredt stratospheric bomber

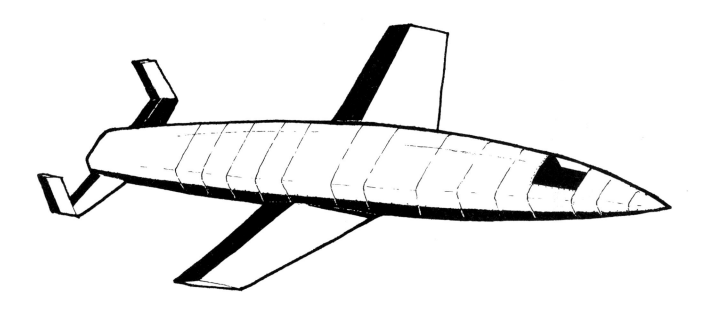

In 1933, at the age of only 28 years, Dr Eugene Sänger published his book *Rocket Flight Technique*, which led eventually to his being invited to set up the German Research Institute for Rocket Flight Technique in 1936. By 1938 Sänger had worked out most of the details of a supersonic glider and had started to build a 1/20th-scale model. But with the start of hostilities in 1939 Sänger was pressured into deciding either to shelve his project or adapt it for military use. The result was a revolutionary design for a single-seat bomber with a fuselage of flat-sided rectangular section and measuring 91ft 10¾in long, 11ft 9¾in wide and 6ft 11in deep (28m × 3.60m × 2.10m).

It was designed for stratospheric flight, with the pilot housed in a forward pressurised cabin. The pointed fuselage opened out in section towards the blunt tail, from which protruded the primary rocket nozzle flanked by a pair of auxiliary rockets. The flattened fuselage created body lift which was supplemented by a pair of comparatively small 49ft 2½in (15.0m) span wings set halfway along the fuselage and blending into the totally flat underfuselage. The horizontal tail surfaces carried small endplate fins.

Although a retractable undercarriage was envisaged, the aircraft was to take off from a streamlined rocket-powered sled set on a monorail track 1.8 miles (3km) in length. The sled was to be powered by an enormously powerful rocket developing 600 tonnes thrust for a maximum of just 11sec. After accelerating along the track the aircraft would climb to over 5,000ft (1,200m) at an angle of 30°, reaching a speed of nearly 1,150mph (1,850km/hr). At that point the aircraft's main powerplant of a single fuselage-mounted 100-tonne-thrust rocket motor would be fired for a maximum of 8min to propel it to a height of over 90 miles (145km) and a speed of 13,730mph (22,100km/hr). Once the main rocket motor had burned out the aircraft would coast at high speed back down to the denser atmosphere at 25 miles (40km, 131,230ft) altitude and then skip back up like a flat stone on water. This manoeuvre would be repeated until momentum had been lost, and the aircraft would then glide down to a normal wheeled landing, having covered a predicted range of up to 14,600 miles (23,500km). The problems of superheating likely to be encountered at such speeds were expected to be reduced by this method, with the aircraft cooling down on the upward coast after reaching peak temperatures at the lower end of the "skip".

By June 1939 Sänger had advanced his designs to the point of carrying out friction tests, but many of the problems of ultra-high speeds, especially as they affected metal structures, were not adequately understood. Sänger was also working on the 100-tonne-thrust rocket motor, which progressed slowly up to 1941 and achieved some remarkable results, exceeding many of the A-4 (V-2) rocket performances at that time.

But then in 1942 work on the rocket motor was cancelled and Sänger, his assistant, Dr Irene Bredt, and his team were set to work at DFS, the German Research Institute for Gliding. No further development of the stratospheric bomber was requested, although Sänger did manage to publish a secret paper in 1944, writing of the possibility of using it to bomb New York using a single free-falling bomb which was to be carried between the two large mid-fuselage oxidant tanks.

Project Saucer

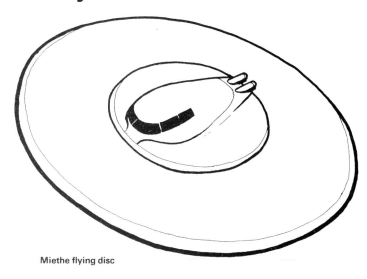

Miethe flying disc

Serious German interest in disc-shaped aircraft seems to have originated from about the spring of 1941, when Rudolf Schriever, a Luftwaffe aeronautical engineer, designed his first "Flying Top", the prototype of which was being test-flown in June 1942. Schriever followed these experiments – carried out in collaboration with three colleagues, Habermohl, Miethe and Bellonzo – by constructing an even larger flying disc in the summer of 1944. It is not clear how these early designs were to be powered, but a bigger version with advanced jet engines was reportedly designed at the BMW factory near Prague.

Information on this aspect of German jet aircraft development is very sketchy. The project was always highly secret, and documents that may have existed were probably either destroyed, lost or taken by the Russians when the war ended. A last possibility is that the Allies discovered Schriever's work and considered it too important to reveal. However, in the late 1950s Rudolf Schriever himself described his work on a wartime research programme named "Project Saucer".

The Schriever and Habermohl design comprised a large-area disc rotating around a central cupola-like cockpit. The disc was made up of adjustable wing surfaces which could be positioned for take-off or level flight. Dr Miethe also developed a 138ft (42m) diameter flying disc powered by vectorable jets. This machine is said in many references actually to have flown, and one source even pinpoints the date of the first flight as February 4, 1945, and the place as Prague. On that flight the machine, designated V-7, is supposed to have climbed to a height of 37,600ft (11,450m) in just three minutes, reaching a level speed of 1,218mph (1,960km/hr). But Schriever claimed after the war that while his flying disc was made ready for testing early in 1945, the preparations were cancelled in the face of the Allied advance, and the machine was destroyed and all information lost or stolen. The factory at Breslau at which Schriever's saucer is said to have been built fell into Russian hands, and the prototype and the technicians working on it are believed to have been captured and taken to Siberia to continue the project under Soviet control. Also rumoured was another near-complete flying disc which was expected to be capable of reaching 3,000mph (4,830km/hr).

Although the evidence for the existence of a German flying-disc programme is very tenuous, the senior official of a 1945 British technical mission revealed that he had discovered German plans for "entirely new and deadly developments in air warfare". These plans must obviously have gone beyond normal jet aircraft designs, as both sides already had jet-powered aircraft in production and operational service by the end of the war. Moreover, before Rudolf Schriever died some 15 years after the war he had become convinced that the large numbers of post-war UFO sightings were evidence

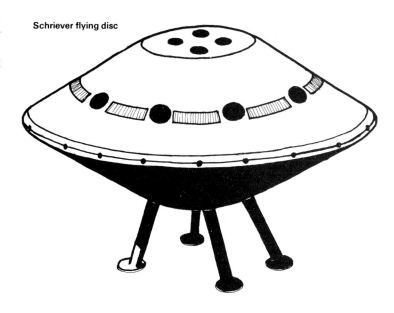

Schriever flying disc

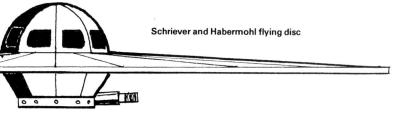

Schriever and Habermohl flying disc

135

that his designs had been built and developed.

The drawings on page 135 show three flying-disc designs attributed to the Schriever group. The first is credited to Dr Miethe and was supposed to have been almost ready for "operational" use when the factory in Prague was captured by the Red Army. A modern set of very basic three-view drawings shows twin cockpits above and below the centre of the disc, each with two jetpipes. The second illustration shows a flying disc, the details of which were revealed in a set of blueprints published in recent years by the West German Government. It has been suggested that the drawings were probably inspired by Rudolf Schriever's own wartime designs but had been made safe for publication by the modification of certain details. The deep disc was nearly 50ft (15m) in diameter and stood on four short legs. Amongst the equipment indicated in the drawings is a laser system, radar, computers and electro-magnetic turbines.

The third drawing depicts a flying disc which is credited in one reference to Schriever and Habermohl, though the claimed details and achievements exactly match those of the Miethe-designed disc which supposedly flew on February 4, 1945.

Skoda-Kauba P.14

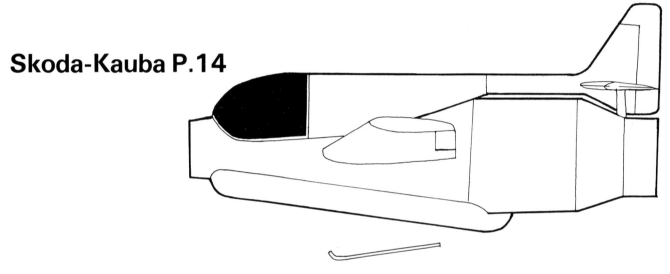

Early in 1945 the Skoda-Kauba Flugzeugbau and DFS began design work on a fighter (see also Heinkel P.1080, page 86) to be powered by a large Sänger ramjet measuring 4ft 11in (1.50m) in diameter and 31ft 2in (9.50m) long. The P.14 was little more than a ramjet with wings, the cockpit and tailplane being mounted above the engine and the pilot lying prone very near the large nose air intake. The meagre armament of a single 30mm cannon was to be mounted above the pilot in the upper portion of the fuselage. The wings were relatively small and unswept, as were the conventional fin and tailplane. Behind the pilot was a large tank containing nearly 300 Imp gal (1,350lit) of petrol-type fuel – or alternatively a mixed load of 154 Imp gal (700lit) and 1,875lb (850kg) of powdered coal – giving a maximum endurance of 43min at 355mph (570km/hr) and 32,800ft (10,000m). The ramjet powerplant necessitated a rocket-boosted take-off from a three-wheel bogie, and there was a retractable skid for landing.

Skoda-Kauba P.14 data

Role	Single-seat ramjet fighter
Ultimate status	Design
Powerplant	One Sänger ramjet, 9,702lb (4,400kg) thrust at sea level and Mach 0.83, 2,820lb (1,280kg) thrust at 32,000ft (10,000m) and Mach 0.815
Maximum speed	621mph (1,000km/hr) at sea level, 545mph at 32,800ft (875km/hr at 10,000m), 535mph at 49,200ft (855km/hr at 15,000m)
Climb rate	12min 42sec to 60,680ft (18,500m)
Range	225 miles at 300mph and 42,650ft (360km at 485km/hr and 13,000m)
Ceiling	60,680ft (18,500m)
Weight	3,256lb (1,480kg) empty, 6,370lb (2,850kg) loaded
Span	25ft 11in (7.9m)
Length	31ft 2in (9.5m)
Wing area	135ft² (12.5m²)
Armament	One MK 103 30mm cannon

Zeppelin Fliegende Panzerfaust

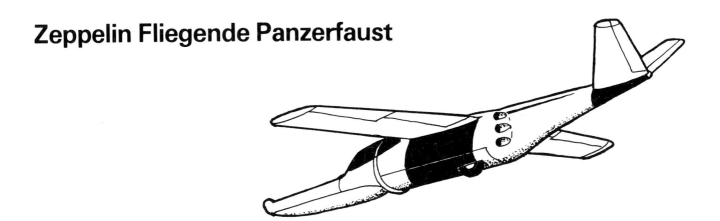

This strange little aircraft was designed as a secondary interceptor for use against bombers. A rocket-powered glider, it was to be towed to within range of the target behind a Bf 109G fighter and then released to attack an individual bomber, leaving the Bf 109 to perform its role as a fighter.

The flying surfaces comprised straight, high-shoulder wings and a butterfly tail, and the pilot lay prone behind a long, hooked nose. This was in fact the towing apparatus, the upturned tip of which was attached directly beneath the tailplane of the Bf 109. Take-off was to be made on a single central wheel semi-submerged in the belly of the aircraft.

After reaching operational height the little interceptor could be cast off, leaving its pilot free to select and attack a target under the power of six solid-fuel rockets fitted three in each side of the fuselage. The endurance of the rockets was just enough to allow the aircraft to reach attacking speed, whereupon the pilot would fire its entire armament of two RZ 65 rocket missiles at the target and retire immediately. Frontal area would have been very small, making the *Fliegende Panzerfaust* (Flying Mailed Fist) a very difficult target for air gunners.

On completion of a mission the pilot and the whole front end of the aircraft could be ejected, parachuting to safety; the remainder of the aircraft would also descend by parachute in a similar manner. The aircraft would then be recovered by a special truck and a team of three men, who would transport it back to the base for reassembly and re-use.

Zeppelin Fliegende Panzerfaust data

Role	Single-seat rocket-powered interceptor glider
Ultimate status	Design
Powerplant	Six small solid-fuel rocket motors
Maximum speed	528mph (850km/hr)
Weight	2,646lb (1,200kg) loaded
Span	14ft 8½in (4.50m)
Length	19ft 8¼in (6.0m)
Wing area	40.9ft² (3.80m²)
Armament	Two RZ 65 rocket missiles

Zeppelin Rammer

There exist drawings of a small rocket-powered aircraft designed for ramming attacks on enemy aircraft in a manner similar to that of the Northrop XB-79. It featured straight mid-wings and a conventional cabin housing a prone pilot and heavily protected with 20-28mm armour plating and 80mm bulletproof glass.

The nose was very long and tubular and was to carry 14 rocket projectiles, fired in a manner similar to that of the armament of the Bachem *Natter*.

It is thought that the aircraft was to be carried or towed by a piston-engined fighter to the required height, where the fuselage-mounted 2,205lb (1,000kg) thrust solid-fuel rocket unit would be fired to gain speed as the aircraft went into attack a bomber formation. After discharging its missiles the pilot would then select and ram a victim, using the reinforced leading edges of the wings to chop off its tail unit. Landings were to be made on a retractable skid.

Zeppelin Rammer data

Role	Rocket-powered single-seat ramming interceptor
Ultimate status	Design
Powerplant	One Schmidding 109-533 solid-fuel rocket motor, 2,205lb (1,000kg) thrust
Maximum speed	600mph (970km/hr)
Span	16ft 5in (5.0m)
Length	16ft 1in (4.90m)
Wing area	62ft^2 (5.75m^2)
Armament	14 R4M 55mm rocket projectiles

Appendix

The last throw...

The following extracts from the Royal Aircraft Establishment report *German Aircraft: New and Projected Types* of January 1946 was based on the minutes of meetings between some of the highest authorities in the wartime German aircraft production and procurement apparatus. Reproduced here verbatim, it gives a good idea of what the Allied air forces might have faced if the war had gone on beyond 1945.

"The minutes of a meeting dated 21/22nd November, 1944, indicate the scope of the development programme decided upon a few months before Germany's defeat. The highest priority was to be given to four key types – He 162, Me 262, Ar 234 and Do 535.

Target-Defence Aircraft
1. The importance of target defence was emphasised and consideration was narrowed down to the 8-248 (8-263), a development of the Me 163B; the Heinkel "Julia"; Bachem "Natter"; and the Me 262 interceptor with supplementary rocket propulsion. It was decided that since these developments were in an advanced state it was not expedient to abandon any of them. A proposal by the Special Commission for Jet Aircraft and Special Aircraft to defer or reject the 8-263 in favour of the He 162 was opposed on the ground that further development and series production of the 263 could be based on the work already undertaken in connection with the 163. The four types of target-defence aircraft enumerated were to be developed in the following priority:
    i) Me 262 with supplementary rocket propulsion
    ii) Heinkel "Julia"
    iii) 8-263
    iv) Bachem "Natter"

The development of the BMW rocket 109.709, using nitric acid, was to proceed on a high priority as this unit was intended for the three last-named developments.

Multi-engined Aircraft
2. The Do 335 was to be further developed as a "bombing-leader" aircraft, but cancellation of the development of the Do 435 was recommended since a suitable makeshift two-seat version of the 335 was available, and the Do 435 represented practically a new aircraft.

Flying Wings
3. The Ho 229 was to be developed in conjunction with Gotha, and three prototypes of the Horten VII were to be completed. The Lippisch P.11, a parallel development with the Ho 229, was to be developed in collaboration with Henschel.

Research Aircraft
4. Emphasis was placed on the athodyd propulsion system of the Lippisch P.13. Concerning the DFS 228 rocket-propelled reconnaissance aircraft, described by the Germans as a "glider rocket unit for altitudes in excess of 65,000ft", it was stated that the best employment of this type would be decided when test results were to hand. Many points were to be clarified: for example, baling out from great heights. Ten prototypes would be completed.
5. Three examples of the 8-332 rocket-propelled glider were ordered. This was a pure research aircraft for profile measurements at high Reynolds numbers which could not be obtained in the wind tunnel.
6. Work on the "1068" piloted flying model with rocket propulsion was to continue as planned. (NB this appears to have been a flying scale model of the He 343.)
7. No firm had been designated to build the 8-346 research aircraft. (NB later Siebel were entrusted with the task. The aircraft was to be used for measurements in the sonic and supersonic ranges.)

Miscellaneous Jet-propelled Aircraft
8. The Junkers EF.126 project, described in the present report as a ground-attack aircraft, was mentioned at the meeting as a heavy fighter with one or two Argus impulse duct units. Its future depended on Junkers' production capacity.
9. It was clear from the minutes of the meeting that the development of the Ju 287 multi-jet high-speed bomber hung in the balance.

10. The Hs 132 dive-bomber, with pilot in the prone position, was to be subject to a decision of the quantity to be produced. This type is also referred to as a fighter.

Helicopters

11. The removal of personnel for Me 262 construction would not mean the abandonment of the Fa 223, Fl 282 and Fa 339 helicopters. Development of the MR 54 ("Knapsack" helicopter) would continue for study of the single-rotor principle; work on the WNF (Doblhoff) helicopter would likewise proceed.

Gliders

12. The He 162S, a training version of the 162 fighter, is mentioned under the heading "gliders" and is followed by a reference to the Re 5, also described as a glider for 162 training. The 162S was to be developed by Heinkel in conjunction with NSFK and the Re 5 by Segelflug-Reichenberg.

13. A second document date-lined Berlin 22nd December, 1944, throws more light on the ideas of German designers and the problems which confronted them. The paper was addressed to the Chief of Technical Air Equipment (*Chef TLR*) and was issued by the Chief Commission for Aircraft Development (*Entwicklungs Hauptkommission*).

14. The following list of special Commissions is given:

Commission	Director
Special Commission for Day Fighters	Messerschmitt
Special Commission for Night Fighters	Tank
Special Commission for Bombers	Hertel
Special Commission for Training Aircraft	Fecher
Special Commission for Special Aircraft	Lusser
Special Commission for Airframe Construction	Bock
Special Commission for Power Units	Schilo
Special Commission for Planning and Installation of Equipment	Stussel
Special Commission for Planning and Installation of Armament	Blume

Single-Seat Fighters

15. Speeches were made by representatives of various firms. Messerschmitt discussed "Otto-Engine and Turbo-Jet Fighters or only Turbo-Jet Fighters". He said that there was complete and unequivocal agreement that in competing with foreign aircraft only the turbo-jet fighter could be decisive. It was known, he said, that turbo-jet fighter development abroad was "approximately on the same performance footing as in Germany" and with this in mind the turbo-jet fighter should be developed as rapidly as possible with the HeS 011 unit. The He 162, particularly with the BMW 003 unit, must definitely not be regarded as the perfect solution, but only as a stop-gap.

16. Messerschmitt went on to say that since it was not clear whether the turbo-jet fighter was adequate for all operational tasks envisaged, by reason of its special characteristics, further development of fighters with reciprocating engines could not be discontinued. It was accordingly concluded that development of the Ta 152 should proceed in order to match the performance of foreign aircraft with i/c engines. It was necessary, he said, to arrive at a conclusion as to whether the performance of the Ta 152 could be improved to compare with that predicted for new projects with i/c engines.

Night and Bad Weather Fighters

17. Kurt Tank said it was agreed that the operational qualities of contemporary German bad weather fighters were not satisfactory; moreover the problem of designing aircraft for day fighting "without optical vision" had not been solved. German night fighters in current use and proposed for the near future were far from adequate for dealing with the Mosquito. Night fighter developments of the Ar 234 and Do 335 were only makeshift solutions and did not satisfy the operation demands of longer endurance and adequate facilities for navigation. He concluded that it was necessary to develop as a matter of the greatest urgency a "superlative" fighter with an endurance of 5 to 8 hours, a crew of three, and satisfactory provision for navigation. Tank mentioned a development of the Do 335 employing a turbojet.

Small Aircraft with Argus Tube

18. Hertel stated that Junkers proposed a very cheap and simple single-seater with an Argus tube power unit (109.014), intended for ground attack operations. (NB this aircraft is described in the present report as the EF 126.) The GAF considered the range of 280-310 miles and the endurance of 40 to 50 minutes to be completely inadequate, so that the realisation of this project was considered as impracticable. However, no objection could be taken to a development of this very interesting and economical aircraft for experimental purposes, but with no intention of quantity production.

Target-Defence Interceptors

19. Reference was made to the inadequate endurance of the Me 163B; and it was further stated that the Heinkel "Julia" and Bachem "Natter" projects did not hold promise. It was concluded that the development of the 263 should be expedited by all means and that the tests of the Me 262 with supplementary rocket propulsion should be pursued as with good results this aircraft might render all other target-defence interceptors superfluous. The development of "Julia" was to be discontinued because of its inadequate endurance. Work on 'Walli" would also be suspended and resumption of development would be dependent upon results obtained with the Me 163, 263 and 262. (NB 'Walli" is the Junkers project EF.127.)

20. Although the "Natter" project was opposed on technical and tactical grounds, the completion of development was agreed to because the initial firing tests

were due soon to take place. All preparations for series production, however, were to be discontinued.

21. Reference was made to a proposal by Willi Messerschmitt to attach an additional rocket unit with special external tanks to existing fighters with internal combustion engines and turbo-jet units. This scheme, it was agreed, should be thoroughly investigated by the firm of Messerschmitt as positive results would provide the simple solution to the problem of target defence.

Engines for Future Aircraft

22. A list of new aircraft actually on test at Rechlin (Germany's Boscombe Down) during February this year is given in a third document headed "Notes on Rechlin Emergency Testing Programmes, 6.2.45".

23. Me 262 production aircraft were being subjected to performance checks and final tests of the Jumo 004 jet-unit installation. Experiments were also in progress with a parachute tail brake, drop tanks and new sub-types. The Ar 234B was undergoing consumption tests and RT installation trials; the sub-type "S" was doing its prototype trials. Prototype trials were all in progress on the Ta 152H and Ta 152C while the Fw 190 was undergoing performance tests with MW 50 and with certain aerodynamic improvements incorporated by Focke-Wulf. Production Me 109s were still being tested and prototype trials of "Jagersteuerung" (FuG 125) were in progress. The He 162 was doing its prototype trials and the Do 335 was finishing RT and engine tests. Some work on pre-spinning of the undercarriage wheels was also in progress. Finally, prototype trials of the Ju 388 with Jumo E engines were in hand."

Bibliography

The contents of this book have been compiled over a number of years from a vast number of major and minor references. This brief bibliography details the most significant of these references, and the author wishes to acknowledge their contribution towards the compiling of this book.

Die deutschen Flugzeuge 1933-1945 by Karlheinz Kens and Heinz Nowarra; J.F. Lehmanns Verlag, Munich

A.I.2(G) Report No 2383

German Aircraft: New and Projected Types; Royal Aircraft Establishment

Warplanes of the Third Reich by William Green; Macdonald and Jane's, London

Warplanes of the Second World War Volume I by William Green; Macdonald & Co, London

Jet Fighters and Bombers by David A. Anderton; Phoebus Publishing, London

Aircraft of World War II by K.G. Munson; Ian Allan Ltd, London

RAF Flying Review

Air Reserve Gazette/Air Training Corps Gazette; Rolls House Publishing Co Ltd